Convair
B-36 Peacemaker
A Photo Chronicle

Meyers K. Jacobsen

Schiffer Military History
Atglen, PA

ACKNOWLEDGMENTS

The author wishes to thank the many individuals and organizations that made this publication possible. The assistance of General Dynamics Convair Division, the San Diego Aerospace Museum, the 7th Bomb Wing B-36 Association, C. Roger Cripliver, James H. Farmer, Frank Kleinwechter, Scott Deaver, Robert W. Hickl, Tim Timmerman and David Menard is gratefully appreciated. Also, previous research work by Lindsay Peacock and Joe Baugher is hereby acknowledged.

DEDICATION

This book is respectfully dedicated to the men of the Strategic Air Command who maintained and flew the Peacemaker. A special commendation goes to the members of the 7th Bomb Wing—the first to operate B-36s.

ADDITIONAL B-36 INFORMATION

Readers and B-36 bomber enthusiasts can get more detailed background information on the history of the Peacemaker in "Convair B-36, A Comprehensive History of America's Big Stick," also published by Schiffer Publishing Ltd. Call 610-593-1777 for ordering information.

Book Design by Ian Robertson.

Library of Congress Catalog Number: 99-64161.

Printed in China.
ISBN: 0-7643-0974-9

We are interested in hearing from authors with book ideas on related topics.

Published by Schiffer Publishing Ltd.
4880 Lower Valley Road
Atglen, PA 19310 USA
Phone: (610) 593-1777
FAX: (610) 593-2002
E-mail: Schifferbk@aol.com.
Visit our web site at: www.schifferbooks.com
Please write for a free catalog.
This book may be purchased from the publisher.
Please include $3.95 postage.
Try your bookstore first.

In Europe, Schiffer books are distributed by:
Bushwood Books
6 Marksbury Road
Kew Gardens
Surrey TW9 4JF
England
Phone: 44 (0)181 392-8585
FAX: 44 (0)181 392-9876
E-mail: Bushwd@aol.com.

Try your bookstore first.

INTRODUCTION: The Convair B-36 Intercontinental Bomber

It was, and still is, the largest bomber ever to be in service with the United States Air Force.

In today's Air Force, the B-1 and B-2 Stealth bomber outperform the old Peacemaker on every front except that of sheer size and bomb carrying capacity. Neither can match the 86,000 pound weapons load that could be carried internally by the B-36. Unrefueled, neither can match the long distance range of the B-36, for it was the first true intercontinental bomber.

During the late 1940s and early 1950s, the B-36 was the mainstay of the USAF's Strategic Air Command, which provided the nation's primary nuclear deterrent. Although the big bomber never dropped a bomb in anger and never saw combat, it did its job well, helping to keep the peace during this early Cold War period. It truly earned its nickname, "Peacemaker."

The story of the Peacemaker begins in 1941, before America had entered the Second World War.

The Nazis were overrunning Europe, and the U.S. was concerned that Britain might fall and leave the Army Air Corps without potential advance bases. Without such bases, a bomber of unprecedented range would be needed to operate from the North American continent. Thus, the need and origin of the B-36 was born.

XB-36 PROTOTYPE

President Franklin D. Roosevelt in conference with Army Chief of Staff George C. Marshall and Maj. Gen. Henry "Hap" Arnold, Air Corps chief, directed that an intercontinental bomber be developed. A design competition was announced by the Army Air Corps on April 11, 1941, that specified, after revisions, that the bomber be able to carry a 10,000 lb. bombload to a target 5,000 miles and return. It also was to have a speed of 240 to 300 mph and operate from a runway of 5,000 ft in length. Service ceiling was to be 40,000 ft. This was a tremendous task for the companies that entered the competition—Consolidated, Boeing, Northrop, and Douglas. The Boeing B-17, then in service, could only carry a 4,000 lb. bombload 1,000 miles and back.

Consolidated Aircraft Corporation of San Diego, California, won the competition with its Model 35 (later changed to 36), a six engine pusher design, and a contract for two prototypes was awarded on November 15, 1941. The first prototype was to be delivered in May 1944, the second six months afterwards. Cost of the two experimental planes including engineering and construction was $15 million, with Consolidated to receive a fixed-fee of $800,000.

In August 1942, the company moved the XB-36 project, including all the engineering drawings, the wooden mock-up, the engineers themselves, and some tooling to Fort Worth, Texas, setting up at the new Government Plant #4. Consolidated had leased the huge factory and was busy building B-24 Liberators for the war effort. Progress was slow on the XB-36 Project, as priority was given to B-24 production and B-32 Dominator development programs.

Consolidated Aircraft Corporation became Convair (Consolidated Vultee Aircraft Corporation) when it merged with Vultee Aircraft on March 17, 1943. At this time, China appeared near collapse against the invading Japanese, and the United States Army Air Forces was concerned about the possible loss of bases in China from which it intended to launch B-29 raids against Japan. The longer-ranged B-36 might be the only airplane capable of attacking the Japanese homelands.

The president of Convair complained to the USAAF that it was difficult to secure subcontractors for an order of only two aircraft, and that the company would be in a better position if there was the promise of a large-scale production contract. Consequently, a "letter of intent" for 100 B-36 bombers was issued on July 23, 1943. Under the new schedule, the XB-36 prototype was to be ready to fly in September 1944. The first production B-36 would be delivered in August 1945, with the last one in October 1946.

Basic configuration of the B-36 changed little over the years from the XB-36 design. Wing span remained 230 ft. with a wing area of 4,772 sq. ft. Length of the fuselage was 163 ft. (162.1 ft. on production models), with a four section sliding panel bomb bay. An 85 ft. pressurized tunnel through the bomb bays connected the forward and aft crew compartments by means of a small trolley. The XB-36 was to be powered by six of the new Pratt & Whitney 28 cylinder R-4360 Wasp Major engines. Each of the 3,000 hp air-cooled radials drove a 19 ft. three-bladed Curtiss propeller in pusher configuration. Six intemal fuel tanks with a capacity of 21,116 gallons were incorporated into the wing for long range flights. 1,200 gallons of oil was also carried.

The XB-36 design initially featured a twin tail arrangement until replaced by a single tail in late 1943, which was almost 47 ft. tall. Unique to the type was the "airliner-type" nose design. A giant 110" diameter single tire main landing gear, the largest tire ever manufactured for an airplane, was installed on the XB-36.

Defensive armament planned for the XB-36 was to consist of five huge 37mm cannon and ten .50 cal. guns. The upper and lower forward turrets, each with two 37mm guns, were to be manned by a gunner in a fashion similar to the turrets on a B-17. No armament was ever actually installed on the XB-36 prototype.

Progress on the XB-36 was still slow, and enthusiasm for the project ebbed and flowed with the fortunes of war. However, by mid-1944, the military situation in the Pacific had greatly improved. Island bases had been secured in order for the USAAF to deploy B-29s and strike the Japanese mainland. The B-36 program continued, but no longer had a high priority.

With Nazi Germany's surrender in May 1945, ending the war in Europe, aircraft production in the U.S. was drastically curtailed. However, the 100 B-36 contract remained intact. The Air Force realized the difficulty and human cost of seizing island bases in the Pacific, and this fact convinced the Air Staff that there was still a need for a long range intercontinental bomber.

Furthermore, the advent of the atomic bomb required a long range delivery system capable of reaching targets without the requirement of first obtaining advance forward bases. In August 1945, when the war ended in the Pacific, the Air Staff recommended that four B-36 groups be included in the postwar USAAF.

Construction continued on the XB-36 prototype even after Japan surrended on September 2, 1945, VJ Day. Labor strikes at Fort Worth in October 1945 and February 1946 resulted in delays beyond what was already being experienced from inadequate materials and poor workmanship. The XB-36 was over two years behind its original schedule.

Six days after VJ Day, the XB-36, 42-13570, was rolled out of the Convair Experimental Building in Fort Worth. It sat on its huge single tire mainwheels, which restricted it to only three airfields in the entire United States that had sufficient runway reinforcement. Taxi tests began on July 21, 1946, and the test pilot, Beryl A. Erickson, was finally ready for its maiden flight. At 10:10 AM on August 8, 1946, the XB-36 lifted off the runway at Fort Worth for the frst time. It was a wheels-down flight, and Erickson, with his eight man crew, cautiously flew it for an uneventful 37 minutes.

However, the XB-36 did display some problems that were troublesome—most of which were eventually resolved. Engine cooling needed improvement, and propeller vibration adversely affected the wing structure. The aircraft's overall performance also fell below expectations, especially in regard to speed. The XB-36 prototype performance record in 1947 included a top speed of 315 mph at 30,000 ft. and a service ceiling of 38,200 ft. Range was estimated to be 9,360 miles with a 10,000 lb. bombload. Gross weight of the XB-36 was 270,000 lbs.

After the second prototype, designated the YB-36, became available in late 1947 for the flight test program, the XB-36 was returned to shop for some modernization, including installation of the new four wheel bogie landing gear of the production models. Convair pilots made 53 test flights with the XB-36, logging a total of 117 flying hours. It was then turned over to Air Materiel Command at Wright Field, Ohio, in June 1948, but later returned to Fort Worth where it was used for a short time for training purposes at Carswell AFB, across the field from the Convair plant.

Since it had limited operational value, consideration was given to modifying the prototype to production B-36 standards. This was determined to be too expensive, and the XB-36 ended its career in 1957 as a derelict used for firefighting training at Carswell.

YB-36

The YB-36, 42-13571, flew for the first time on December 4, 1947. It featured a new high-visibility glass canopy over a redesigned crew compartment. This improved design would enable installation of nose armament in the production models. However, the YB-36 still shared the same huge single wheel landing gear of the XB-36. As for performance, it reached an altitude of 40,000 ft. during its third flight. GE BH-2 turbosuperchargers being installed in the production models helped the YB-36 to easily outperform the XB-36.

B-36 PROGRAM DEBATE

On December 12, 1946, Gen. George S. Kenney, SAC's first commander, suggested that the procurement contract for 100 B-36s be reduced to only a few service test airplanes. He believed the B-36 to be inferior to the Boeing B-50, an improved version of the B-29. Shortcomings of the B-36 were stated to be an effective range of only 6,500 miles, insufficient speed and lack of protection for the fuel load. However, the Air Staff and Gen. Nathan F. Twining, commander of AMC, disagreed with Kenney's assessment and felt the problems being experienced by the B-36 at this stage in its development were normal and could eventually be solved. Gen. Carl A. Spaatz, commander of USAAF, agreed with Gen. Twining, and the B-36 contract was retained.

In fall 1947, the new USAF Aircraft and Weapons Board held a conference to determine which aircraft would best support the Air Force's long term plans. At that time, the B-36 was the only bomber capable of carrying atomic weapons against an enemy without the need for overseas bases. Some members of the Board felt the B-36 was obsolete and should be canceled in favor of fast jet bombers, such as the B-47. After prolonged debate, it was decided to keep the B-36 as a special purpose nuclear deterrent bomber. It was thought at the time that 100 B-36s would be enough, and no further production was planned beyond the original contract.

B-36A

The initial production version was the B-36A. The first airplane of the series off the assembly line was B-36A, 44-92004. It flew for the first time on August 28,

1947, actually four months before the YB-36 took to the air. It carried no armament and only enough equipment for a one time flight to Wright Field, where it was gradually destroyed during static structural testing.

Another 21 B-36As were built by Convair (44-92005/44-92025). The first four B-36As to join the Strategic Air Command at Carswell AFB were 006, 007, 015 and 017. 015 was the first airplane delivered to the 7th Bomb Group (Heavy) in a ceremony on June 26, 1948. None of the B-36As had any armament and were used solely for training and crew familiarization. The B-36A model had the same R-4360-25 3,000 hp Wasp Majors as the two prototypes and could achieve a maximum speed of 345 mph at 31,600 ft. Cruising speed was 218 mph, service ceiling 39,100 ft. Combat radius was 3,880 miles with a 10,000 pound bombload. Maximum gross weight was 310,380 lbs.

B-36B

The first fully equipped combat model of the B-36 was the B-36B. It differed from the B-36A in having 3,500 hp R-4360-41 Wasp Major engines with water injection. Having an additional 500 hp from each of the six engines enabled the B-36B to take off from a shorter runway and yield somewhat better performance at both maximum and cruising speeds.

The B-36B had upgraded electronic equipment, including the AN/APQ-24 bombing/navigation radar. It could carry a maximum load of 72,000 pounds of bombs and was equipped from the beginning with six remote-controlled retractable turrets, each with a pair of 20mm cannon, plus two more 20mm cannon each in the nose and tail turrets. The crew of the B-36B was normally fifteen—a pilot, co-pilot, radar operator/bombardier, navigator, flight engineer, two radiomen, three forward gunners, and five rear gunners.

The first B-36B flew on July 8, 1948, with better performance than the B-36A. Top speed was 381 mph as compared to the 345 mph of the previous model. The first B-36Bs were assigned to the 7th Bomb Group in November 1948, and its B-36As were gradually transferred to the newly forming 11th Bomb Group, also at Carswell.

On December 5, 1948, a 14 hour long range mission of 4,275 miles was flown at 40,000 ft. Also, on December 7-8th, the anniversary of the attack on Pearl Harbor, a B-36B flew a 35 1/2 hour simulated combat mission from Texas to Hawaii and back. Carrying a dummy 10,000 lb. bomb, which was dropped just off Honolulu, the undetected mock attack was an embarassment for Hawaii defense officials. Total distance flown was in excess of 8,000 miles. On January 26, 1949, a B-36B established a record bomb lift by carrying a pair of dummy 42,000 lb. "Grand Slam" bombs aloft at Muroc (later Edwards) AFB. The first was released at 35,000 ft., the second from 40,000 ft. In March 1949, another Carswell B-36B set a new long distance record of 9,600 miles on a flight that lasted 43 hours, 37 minutes.

The B-36B was the first B-36 modified to carry the early atomic weapons. None of the B-36As had been configured to handle atomic bombs, largely because engineering specifications pertaining to the atomic bomb had been withheld from Convair for security reasons. Weapons that armed later B-36s included the huge Mk 17 thermonuclear device, which weighed 21 tons. Other weapons were the Mk III, Mk IV, Mk 5, Mk 6, Mk 15, Mk 18, and Mk 36.

Although the B-36B flew a series of impressive demonstration flights, teething problems were evident from 7th Bomb Group evaluations. The remote-controlled turrets and 20mm guns were quite complex and prone to frequent failures. Parts shortages were acute, and it was necessary to cannibalize some B-36Bs just to keep others in the air. Ground equipment such as stands, dollies, and jacks were also in short supply. In reality, it would not be until 1952 that full operational capability would be achieved.

The B-36B had a maximum speed of 381 mph at 34,500 ft; cruising speed of 212 mph; service ceiling of 42,500 ft. with a combat ceiling of 38,800 ft; range of 8,175 miles; and a gross weight of 328,000 pounds. Of the 62 B-36Bs built, 59 were later converted to B-36D configuration, from 1950 to 1952. All but five of these conversions were at Convair's San Diego plant.

B-36C

In order to increase speed, Convair proposed in March 1947 that 34 of the 100 B-36s be fitted with a proposed new version of the R-4360 Wasp Major, called the Variable Discharge Turbine, or VDT. It would require redesigning the engine installations on the B-36B to forward pulling, or tractor propellers. Convair claimed the VDT engine would give the B-36 a top speed of 410 mph, a 45,000 service ceiling, and a 10,000 mile range with a 10,000 pound bombload.

Convair proposed the last 34 B-36s in the 100 airplane contract be completed as B-36Cs, with the extra cost being met by reducing the original contract to 95 airplanes. Unfortunately, the VDT B-36C project ran into many technical difficulties related to mating the tractor version to the B-36's wide wing. By the spring of 1948 it was apparent that the VDT engine adapted to the B-36 was not going to materialize. With the higher performance B-36C canceled, the Air Force considered once again whether to cancel the entire B-36 program.

However, tests had shown that the B-36B surpassed the B-50 in cruising speed at long range, had a higher altitude, carried a larger bombload, and had a much greater combat radius than the B-50.

It now seemed the B-36 might be a better bomber than anyone had expected. World events then played a role in saving the B-36 program.

The Soviet's blockade of the city of Berlin began on June 18, 1948. Cold War tensions were high. Urgency was now a factor in securing a strategic bombing force, and Air Force Secretary W. Stuart Symington decided to stay with the B-36 program since it was the only true intercontinental bomber then available (In-flight refueling was not yet fully developed).

Gen. Kenney, SAC's commander, agreed—even though he had earlier criticized the B-36, favoring instead the B-50. The proposed 34 B-36Cs would finally be completed as B-36Bs. Later, with in-flight refueling perfected, the B-50 would join the B-36 in SAC for a number of years during the late 1940s to mid-1950s

B-36 CONGRESSIONAL HEARINGS

The United States Army Air Forces had become a separate military service from the Army on September 18, 1947. When B-36Bs started entering the SAC inventory in the fall of 1948, the newly independent U.S. Air Force had 59 groups. The USAF wanted to expand to 70 groups, but was thwarted by Fiscal Year 1949 budget restraints. President Harry S. Truman was determined to hold the FY 49 defense budget to $11 billion. The three military services squabbled with each other over who was to receive the lion's share of the money. The Air Force wanted more B-36s, but the Navy wanted a new supercarrier, the first of four, that would give them a strategic bombing capability. The Air Force's position was that strategic bombing should remain an Air Force responsibility, and that a Navy strategic bombing capability was redundant.

Gen. Curtis E. LeMay, who had taken over command of SAC in October 1948, recommended that more B-36s be acquired and B-52 production be stepped up. The stage was set for one of the ugliest and most bitter interservice confrontations in U.S. military history.

A lot of criticism was being fired against the B-36. During 1948, rumors circulated that undue favoritism and corruption had entered into the award of the B-36 contract and that the performance of the B-36 did not live up to Air Force claims. The B-36 became the center of a heated political controversy. At this time, the Secretary of Defense was Louis A. Johnson, who had replaced James V. Forrestal on March 28, 1949. Johnson once was on the Board of Directors of Convair. Fireworks began when Secretary Johnson abruptly canceled the first supercarrier, the "USS United States," then under construction, and proceeded with plans to purchase more B-36 bombers for the Air Force. The decision had been made on the grounds that with budget limitations, the government could not afford both new strategic bombers and a new carrier force.

On May Day, 1949, the Soviets showed off a new swept-wing jet interceptor, the MiG 15, and there were doubts the B-36 could successfully defend itself against this fast new interceptor. Individuals, particularly those in the Navy, expressed concerns that the Air Force was spending a fortune on what could turn out to be a "sitting duck."

In August, an anonymous report circulated around Washington that accused the Air Force of grossly exaggerating the importance of strategic warfare. Finally, the House Armed Services Committee launched an investigation on what became known as the "B-36 Controversy." After several weeks of hearings and the testimony of Floyd B. Odlum, Chairman of the Board of Convair, Air Force generals George Kenney and Curtis LeMay, and Secretary of the Air Force Symington, the investigation closed down after clearing both the Air Force and Convair of any impropriety.

The B-36 congressional hearings resumed in October, this time to debate whether the defense of America should rely on a fleet of bombers or on the Navy's proposed fleet of supercarriers. The Navy was still enraged at the cancellation of its first supercarrier. The Secretary of the Navy, John L. Sullivan, had even resigned in protest over the action.

A parade of famous generals, admirals, government officials, and others appeared before the Committee and gave their testimony and opinions. Admiral Arthur W. Radford, commander of the Pacific Fleet, denounced the B-36 as a "billion dollar blunder." Although there were still doubts about the B-36's ability to survive enemy fighter attacks, the Air Force's B-36 program survived the Navy's salvos, and production continued on the Peacemaker.

B/RB-36D

The B-36B had been berated for being too slow during the congressional hearings, and Convair had been busy working on ideas to increase the plane's speed. The VDT tractor engine concept had failed, and an earlier study to equip the B-36B with four tractor and four pusher turboprop engines mounted in tandem never became a reality. On October 5, 1948, at the same time as the B-36 congressicnal hearings, Convair proposed to the Air Force that two pairs of turbojets in pods be installed underneath the outer wing panels. The engines used would be General Electric J47-GE-I9 turbojets of 5,200 lbs. of static thrust each, the same basic engines being used to power the Boeing B-47. Development time was saved by also using the same engine nacelle as the Stratojet. The four turbojets would be used for take off and for when short bursts of power were needed for climbing or dashing over a target area.

The addition of turbojets resulted in the B-36D model. With the jets assisting the six piston R-4360-41 engines, maximum speed was increased to over 400 mph, though cruising speed, without the jets, was 212 mph. Service ceiling was improved to 43,800 ft., and take off run was reduced by almost 2,000 ft. Other improvements included quick-action, split bomb bay doors and metal-covered control surfaces. The B-36D had a better bombing and navigation system, the K-3A, that replaced the B-36B's APG-24 radar. Take off and landing weights were increased to 370,000 lbs. and 357,000 lbs., respectively.

The prototype B-36D was a converted B-36B, 44-92057, and it flew first on March 26, 1949, with Allison

J35s, since J47s were not yet available. The modification proved successful, and the prototype B-36D demonstrated a speed of 400 mph at 38,280 ft. and reached an altitude of 40,000 ft. Anticipation of improved performance led the Air Force, in January 1949, to decide to buy 39 more B-36s as bombers and convert the unarmed B-36As to RB-36E reconnaissance models.

The first true B-36D flew on July 11, 1949, a month before the congressional hearings had started. A year later, 26 jet-augmented B-36Ds had been converted from B-36Bs and delivered to SAC. On January 16, 1951, six B-36Ds were flown from Carswell AFB in Texas to England, landing at RAF Lakenheath after having staged through Limestone AFB in Maine. The flight returned to Carswell on January 20. This mission demonstrated the global reach of the B-36, as well as being the first time B-36s had flown and landed beyond U.S. territory. Another long distance flight to French Morocco was made on December 3, when six B-36Ds of the 11th Bomb Wing touched down at Sidi Slimane, having flown nonstop from Carswell AFB.

As more B-36s were delivered to the 7th and 11th Bomb Wings at Carswell and the 28th Strategic Reconnaissance Wing at Rapid City AFB (later Ellsworth AFB) in South Dakota, most of the mechanical problems with the B-36 were being identified and corrected. An early and major B-36 problem was leakage of the fuel tanks. The electrical system was also unreliable and caused frequent fires. Improved containers and better sealers reduced fuel tank leakages, and changes in the electrical system reduced fire hazards during ground refueling operations.

Landing gear and bulkhead failures were practically eliminated. However, even by October 1951 the B-36's defensive armament system was still operating poorly. In April 1952, Gen. LeMay ordered a series of gunnery tests to see if the cause of the failures could be determined.

The remote control gun system was difficult to operate and maintain, and training for the gunners was found to be inadequate. Tests continued into 1953, and in time these problems were solved.

Several B-36Ds were later modified as featherweight (lighter weight), high altitude aircraft—being stripped of all armament except the tail turret. All nonessential flying and crew comfort equipment was taken out, and the crew was reduced to 13, two fewer than the standard B-36D crew.

26 B-36Ds were built from scratch at Fort Worth, in addition to some 54 B-36Bs that were converted to B-36Ds at Convair's San Diego Lindbergh Field facility. The last B-36D was taken out of SAC service in 1957.

The RB-36D was a specialized reconnaissance version of the B-36D. It was almost outwardly identical to the standard B-36D, except for additional antennas, camera windows, and radomes.

It carried a crew of 22 rather than 15; the additional crew members were needed to operate and maintain the photographic reconnaissance equipment aboard. The forward bomb bay was replaced with a manned, pressurized cabin and filled with fourteen cameras, while the second bay carried 80 T-86 flash bombs, the third an auxiliary 3,000 gallon fuel tank, and the fourth had additional countermeasures equipment (ECM). It did retain all sixteen of its 20mm guns for defensive armament. Performance of the RB-36D and the similar RB-36E conversions was nearly the same as the B-36D, with a maximum speed of 406 mph.

The first RB-36D, 44-92088, made its initial flight, without auxiliary jets, on December 19, 1949—six months after the first B-36D had flown. The RB-36D actually preceded the B-36D into service with the Strategic Air Command by a couple of months. The first unit assigned RB-36Ds was the 28th Strategic Reconnaissance Group at Rapid City AFB. Initially, the group had received 15 B-36Bs for training purposes until its first RB-36D arrived in June 1950. Due to materiel shortages, the new RB-36Ds did not become operationally ready until a year later.

A total of 24 RB-36Ds were built, and ten of these were later converted to GRB-36D FICON parasite fighter carriers. Some RB-36Ds were later modified to featherweight configuration in which all but the tail guns were removed—the 22 man crew was further reduced to 19. These airplanes were redesignated RB-36D-III, with Convair doing the modification work from February to December 1954.

RB-36E

Early in l950, Convair began conversion of the 21 B-36As (and the sole YB-36) to reconnaissance models. These converted planes were redesignated as RB-36Es and were almost identical to RB-36Ds. Their six R-4360-25 Wasp Major engines were replaced with the more powerful 3,500 hp R-4360-41s installed on the B-36B/Ds. They were also equipped with J47 turbojets as fitted on the RB 36Ds. The fourteen cameras in the forward bay included K-17C, K-22A, K-38, and K-40 types. Normal crew was 22, which included five gunners. The last RB-36E of the conversion program was completed in July 1951 and assigned to the 5th Strategic Reconnaissance Group (later Wing) at Travis AFB, California, along with the other RB-36E aircraft.

B/RB-36F

The next model in the B-36 series was the B-36F, and its reconnassaince counterpart, the RB-36F. This improved version had more powerful 3,800 hp R-4360-53 Wasp Majors. Each of these engines generated 300 hp more than those of the B-36D. There was also improved radar and ECM equipment. The first B-36F, 49-2669, took off on its maiden flight on November 18, 1950. The first B-36Fs became operational with

SAC in August 1951. There were some difficulties with the new engines, including excessive torque pressure, ground cooling, and combustion problems. These problems were resolved in a relatively short time.

The K-3A radar system and APG-32 gun laying radar were made standard, and beginning with B-36F 54-1064, a chaff dispenser, was installed in the tail to confuse enemy radar. The last of 34 B-36Fs was manufactured in October 1952, but the Air Force did not get its last one until several months later. A number of B-36Fs were modified as featherweight aircraft during 1954.

The Air Force ordered 24 of the reconnaissance versions, the RB-36F. The first four RB-36Fs were accepted in May of 1951, and the remainder between August and December 1951. The RB-36F's performance was close to the standard B-36F.

Performance of the B-36F included a top speed of 417 mph at 37,100 ft; cruising speed 235 mph; service ceiling 44,000 ft; and a combat ceiling 40,900 ft.

B-36G/YB-60

The B-36G was the initial designation applied to a swept-wing, jet-powered version of the B-36.

Two B-36Fs, 49-2676 and 49-2684, were pulled from the Fort Worth production line and modified as B-36Gs. It was decided to redesignate the two jet bombers as YB-60s, since it was practically a new airplane. However, it still shared 72% parts commonality with the B-36.

Only one of the two experimental prototypes ever flew, making its first flight on April 18, 1952. The other plane never received its J57 jet engines. After losing a production contract to Boeing's B-52 Stratofortress, both YB-60s were scrapped in mid-1954.

B/RB-36H

The B-36H was essentially the same as the B-36F externally, and was powered by the same improved 3,800 hp Wasp Major engines and J47 jets. The H model differed mainly in internal details. A rearranged flight deck with a second flight engineer's station was added. An improved bombing system called Blue Square was installed, and K system components were relocated to a pressurized compartment, enabling access at high altitudes. A new AN/APG-41A radar system aimed the two 20mm cannons in the tail. It was far superior to the AN/APG-32 gun, laying radar used on the preceding B-36Ds and B-36Fs. The new installation featured twin tail radomes.

The B-36H was first flown on April 5, 1952. Deliveries started in December, by which time the Air Force had accepted most of its B-36Fs. 83 B-36Hs and 73 RB-36Hs were delivered from May 1952 to July 1953. A total of 156 B/RB-36s were accepted by the Air Force, making it the largest production run of any B-36 model.

As a test, B-36H, 51-5710 was converted into a probe and drogue mid-air refueling tanker. The Air Force was interested in refueling jet aircraft at higher altitudes and speeds than those reached by KB-29 tankers. The modification contract was approved in February 1952, and tests with a B-47 receiver plane were completed by the end of May. No other tests took place until January 1953, when an improved multiple aircraft system was installed. A nine man tanker crew could convert a standard B-36 into a tanker by installation of a removable 3,000 gallon fuel tank in the bomb bay. The process took just twelve hours. However, no further tanker conversions were carried out, since the new KC-97 could handle mid-air refueling much more economically, and the Air Force felt its B-36s were better utilized in their bombing/reconnaissance roles.

Three B-36Hs, 50-1085/51-5706/51-5710, were also modified by Convair in 1952 to test the Bell GAM-63 Rascal air-to-surface guided missile. It was 31 ft. long, with a launch weight of about 13,000 lbs. At a top speed of Mach 2.95, the missile could carry a 3,000 pound nuclear warhead up to 100 miles. Some eleven other B-36s were scheduled to be modified as Rascal carriers under the designation DB-36H. However, the Air Force decided in 1955 that the B-47, not the B-36, would carry the GAM-63, and the Convair project was eventually canceled. The Rascal program itself was later canceled on September 8, 1958.

The B-36H equipped 42nd Bomb Wing at Loring AFB in Maine was the first unit to start converting to the new all-jet B-52 in June 1956.

Performance of the B-36H included a top speed of 416 mph at 31,120 ft, cruising speed 234 mph, and a service ceiling of 44,000 ft. Maximum gross weight was 370,000 pounds, combat weight 253,900 pounds. Featherweighted B-36F and B-36H bombers are credited with a top speed of 423 mph and a 47,000 service ceiling, the best performance of any B-36 models.

B-36 IN SAC SERVICE

The Strategic Air Command's first B-36 unit was the 7th Bomb Group (H) at Carswell AFB, Texas. It was part of the 8th Air Force and was home to the Air Force's largest bomber at that time, the Boeing B-29 Superfortress. Carswell's location directly across the field from the Convair plant that produced the B-36 was beneficial when trying to resolve early technical problems. The second unit to receive B-36s, the 11th Bomb Group, was also at Carswell, and the two B-36 groups shared the flightline. The third B-36 unit was the 28th Strategic Reconnaissance Group at Rapid City AFB, which became SAC's first RB-36 group after briefly training in bomber versions.

In December 1950, during the first year of the Korean War, SAC had two Heavy Bomb Groups with 36 B-36s and one Heavy Reconnaissance Group with 20 RB-36s. A typical B-36 Bomb Group had 18 aircraft composed of three combat squadrons of six planes. Each unit also had a couple of "spares" assigned. No B-36s were used in Korea, since B-29s were adequate for the task required, and all B-36s remained in the continental U.S. for strategic deterrence.

Originally, March AFB, in southem Califomia, was to be the second RB-36 base, but it received B-47s. Instead, Travis AFB in northern California near San Francisco became the second RB-36 unit, getting its first RB-36 in January 1951.

During 1951, SAC units were reorganized and renamed from groups to wings. The 92nd Bomb Wing at Fairchild AFB in Washington state got its first B-36s on July 29, 1951. Fairchild also became a two B-36 unit base, like Carswell, when the 99th Strategic Reconnaissance Wing joined the 92nd the following month after receiving its first RB-36s. During 1955, the 99th was home to GRB-36D FICON unit, which was teamed with RF-84F/K fighters from the 91st Strategic Reconnaissance Squadron at nearby Larson AFB.

The year 1952 saw B-36s at Walker AFB, New Mexico (6th Bomb Wing), and Ramey AFB in Puerto Rico (72nd Strategic Reconnaissance Wing). Also in 1952, the number of B-36s allocated to operational units was increased to 30 in a bomb wing, with 10 aircraft to a squadron. "Spare" aircraft raised the number in each wing to around 36 total. RB-36 strength peaked by the end of 1953 at 137 airplanes.

The 42nd Bomb Wing at Loring AFB, Maine, became a B-36 unit in April 1953, followed in August by the last B-36 wing to be formed, the 95th Bomb Wing at Biggs AFB, Texas. In August and September 1953, B-36s of the 92nd Bomb Wing completed the first mass flight to the Far East, visiting bases in Japan, Okinawa, and Guam. The flight took place shortly after the hostilities ended in Korea, and was an effort to demonstrate U.S. willingness to maintain operations in Asia. On October 15-16, the 92nd Bomb Wing again left Fairchild AFB and made another long distance flight to the Far East, this time for a 90 day deployment to Guam. It was the first time an entire B-36 wing had been deployed overseas.

By the beginning of 1954, the Air Force's planned force of six Heavy Bomb Wings and four Heavy Reconnaissance Wings was operational at eight bases in SAC. Convair/General Dynamics production of B-36s ended in 1954, but the SAM-SAC program (Specialized Aircraft Maintenance—Strategic Air Command) returned each plane to the factory for equipment updating and maintenance until spring 1957.

SAC's RB-36s never saw combat, but some RB-36s flew rather hazardous reconnaissance missions near or perhaps over Soviet or Chinese Communist territory.

The year of the highest number of B/RB-36s in service was 1954. SAC inventory shows 209 B-36s and 133 RB-36s, a total of 342 aircraft. Also in 1954, filming started on the movie "Strategic Air Command," which starred Jimmy Stewart and June Allyson, and of course, the impressive B-36 bomber. The Paramount picture became the studio's top grosser of 1955. Some of the most beautiful and dramatic aerial sequences ever put on film added to the popularity of "SAC."

The four RB-36 Strategic Reconnaissnace Wings were redesignated Heavy Bomb Wings in October 1955. All RB-36s were converted to bombers, but retained a latent reconnaissance capability. The introduction of new jet reconnaissance aircraft, including the U-2 spy plane, helped seal the fate of RB-36s, which were now obsolescent.

B-36J

The B-36J was the final production vesion of the B-36. Only 33 airplanes were built, and the last came off the assembly line on August 14, 1954. It was the end of an era.

B-36Js featured two additional fuel tanks, one on the outer panel of each wing, which increased the fuel load by 2,770 gallons. It also had a strengthened landing gear, permitting a gross take off weight of 410,000 pounds. The B-36J was first flown on September 3, 1953, with delivery to SAC beginning the following month.

The last 14 B-36Js were manufactured as B-36J Featherweight IIIs with all guns removed, except for the tail position. The crew complement was reduced to 13, and the scanning blisters were replaced by flush covers with small windows. The reduction in weight enabled a service ceiling of 47,000 ft. to be reached, although some missions were evidently flown as high as 50,000 ft. In contrast to other B-36 featherweights which were modified after delivery, these aircraft were built as such on the production line.

Performance of the B-36J included a maximum speed of 411 mph, cruising speed of 203 mph, and a service ceiling of 39,990 ft. Featherweight III B-36Js had a combat range of almost 4,000 miles, maximum speed of 418 mph at 37,500 ft, and a service ceiling of 43,600 ft.

In the mid-1950s, B-36s were gradually being replaced plane by plane with B-52s. Scrapping of the B-36 fleet had begun. Planes were flown directly from their units to Davis Monthan AFB in Arizona, where the Mar-Pak Corporatian handled their reclamation and destruction. In 1956, B-36s from the 42nd Bomb Wing, 92nd Bomb Wing, and 99th Bomb Wing all made their final flights to Arizona. 1957 saw more B-36s from the 6th Bomb Wing, 11th Bomb Wing, and 28th Bomb Wing make the same sad flight, reducing the active force to 127 by year's end. The last full calendar year of operations for the Peacemaker was 1958, when the 5th Bomb Wing at Travis and the 7th Bomb Wing at Carswell consigned their B-36s to the scrap heap.

Defense cutbacks in FY 1958 had forced B-52 procurement to be stretched out, and consequently, the service life of the B-36 had been extended. The remaining operational B-36s were supported by components salvaged from planes already sent to Davis Monthan's boneyard. The 72nd Bomb Wing at Ramey flew its final B-36 mission on New Year's Day 1959. In December 1958, only 22 B-36s were left in the Air Force inventory (all J models).

On February 12, 1959, the last B-36 at Biggs AFB, Texas, was flown to Amon Carter Field in Fort Worth to become a memorial. Within two years, all B-36s had been scrapped except those saved for museum exhibits. The last flight ever of a B-36 took place on April 30, 1959, when B-36J 52-2220 flew north to the Air Force Museum at Wright Patterson AFB, Ohio.

The Air Force accepted a total of 383 B-36s, including the two prototypes, service test aircraft, and reconnaissance aircraft, but not including the two B-36s converted to YB-60s. Average cost of a B-36 was $3,776,000. The entire B-36 fleet was estimated to cost $2 billion in 1950 dollars.

As of summer 1999, four B-36s remain—three currently on public display. A RB-36H, the only RB and H model, is exhibited at the Castle Air Museum in Atwater, California. Two B-36Js are located at museums in Ohio and Nebraska. B-36J 52-2220 can be seen at the Air Force Museum at Wright Patterson AFB in Dayton, Ohio, and B-36J 52-2817 is on display inside the new Strategic Air Command museum near Omaha, Nebraska. A third B-36J, 52-2827, has been restored by volunteers and is presently being stored in several hangars. Plans are to eventually exhibit the plane in a proposed new museum building at Alliance Airport, north of Fort Worth.

Major sections of the YB-36 prototype (in its final configuration as a RB-36E) are stored at a closed to the public private air collection near Cleveland, Ohio.

Artist rendering of proposed Nakajima G10N1 "Mount Fuji" six-engined long range bomber designed to carry out bombing missions against the U.S. from bases in Japan. Similar in size to the XB-36 with a 206 ft. wingspan and 131 ft. length, the bomber was to be powered by 2,500 hp Nakajima NK11A radials. It was to have an 11,000 lb. bombload and a top speed of 423 mph. G10N1 bombers might have been bombing California cities in 1946 or 1947 if the war had proceeded differently. Still on the drawing board at war's end, the Japanese, like the Germans, had realized too late the importance of a strategic bombing force. (Ilustration by John Batchelor)

Another six engine design of the Axis was the Junkers Ju 390. Powered by six 1,700 hp BMW 801D radials, the Ju 390-1 was flown for the first time in August 1943 as an unarmed cargo plane. A maritime reconnaissance version, the Ju 390-2, was a longer aircraft with a range of 6,000 miles and a top speed of 314 mph. Armament consisted of four 20mm cannons and three 13mm guns. It also was equipped with search radar. The second model had a wingspan of 165 ft. and was 121 ft. long, bigger than a B-29. The twin tail plane was delivered in January 1944 to a base near Bordeaux and once flew a 32 hour transatlantic patrol said to have turned back to occupied France only twelve miles off the U.S. coast, just north of New York City. If the Germans had converted the Ju 390 to a bomber, it might have been the first true intercontinental bomber, not the B-36 Peacemaker. (San Diego Aerospace Museum)

Model 36 Design Study in February 1942. (Consolidated Aircraft Corporation)

1/26th scale wind tunnel model of the XB-36. Earlier engine nacelle air intake ducts and original twin tail arrangement are evident. June 1942. (Consolidated Aircraft Corporation)

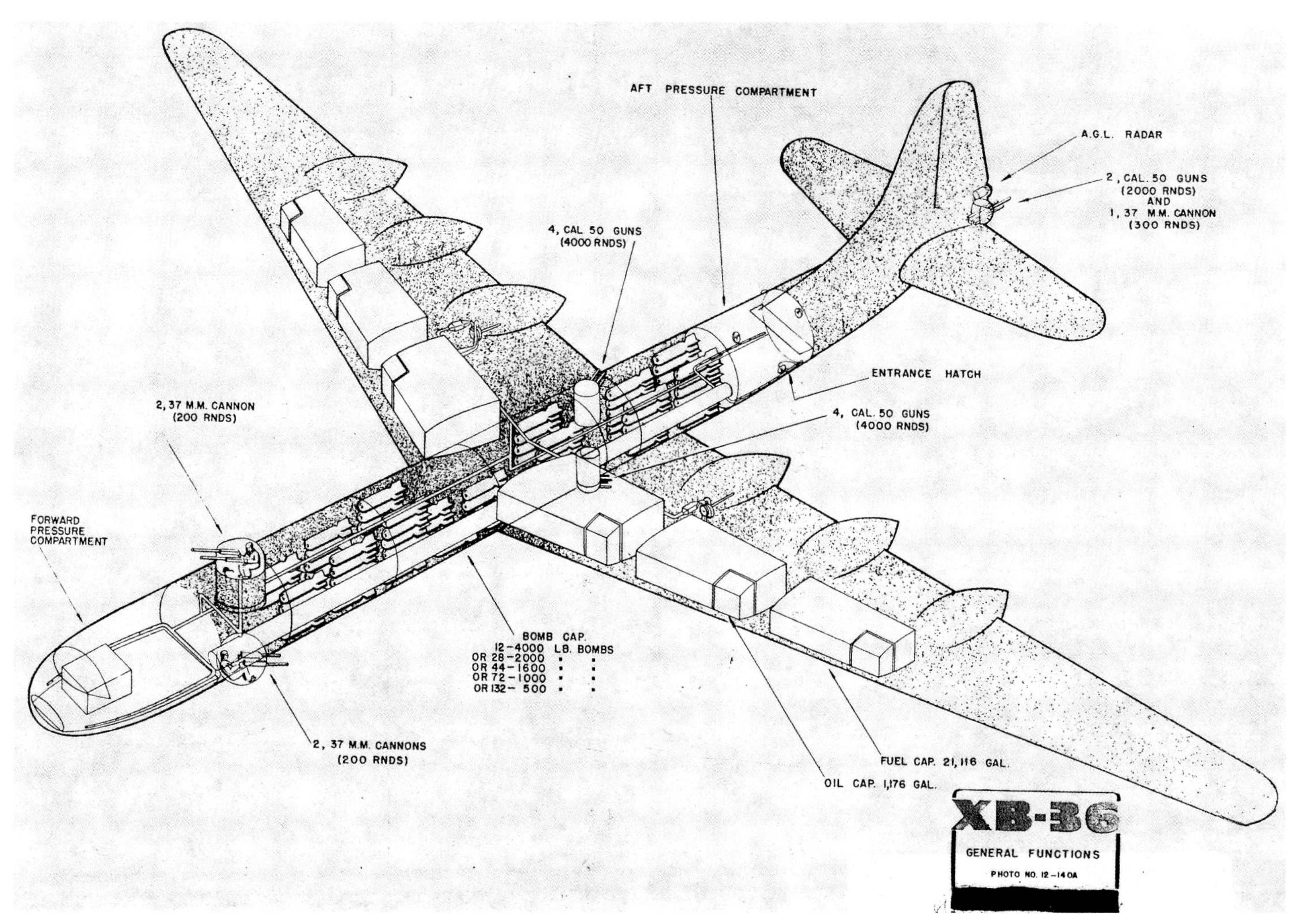

XB-36 general functions showing fuel and oil tanks, bomb bay arrangement, and early armament system. Notice the crewmen in the forward upper and lower turrets, similar to WWII B-17 or B-24 gunners. (Consolidated Aircraft Corporation)

The XB-36 Prototype at rollout, 82.5% completed. Two B-32 Dominators in the background will quickly be scrapped with the end of the war. (Convair)

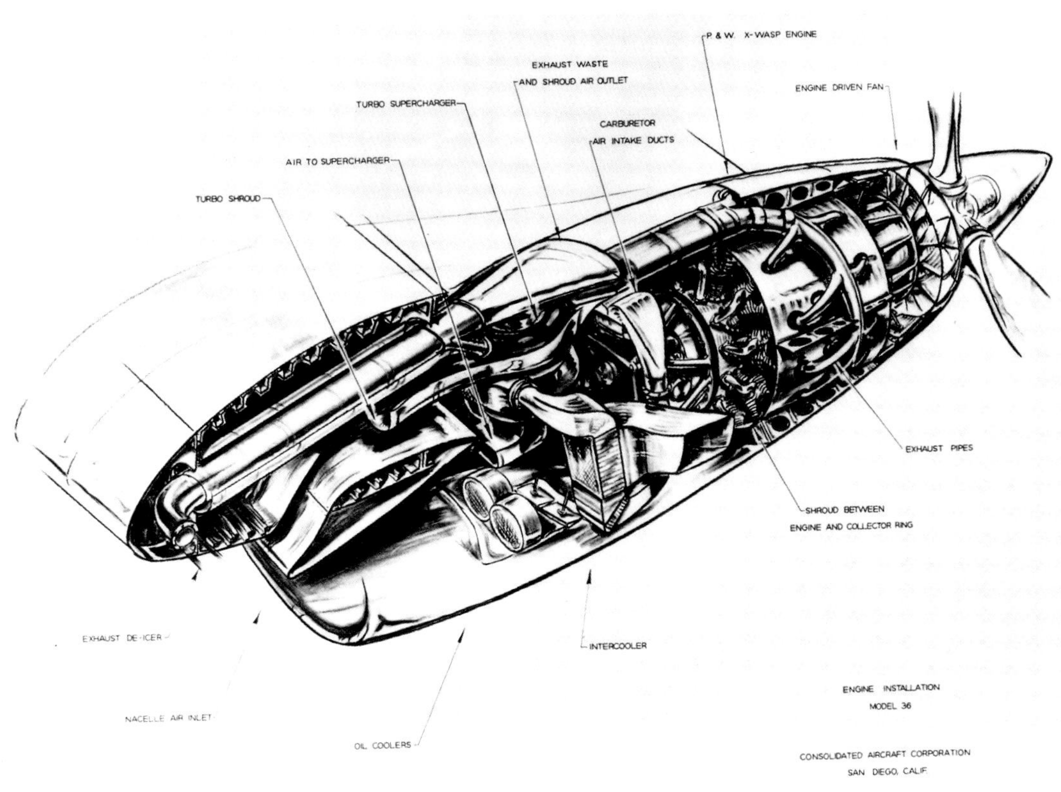

Artist's rendering of B-36 engine nacelle design that was to house the Pratt & Whitney "X" engine that became the 28 cylinder R-4360 Wasp Major. (Convair)

Engine test nacelle outside the Experimental Building at Convair, Fort Worth, where the new 3,000 hp Wasp Major engine was run hundreds of hours. (Convair)

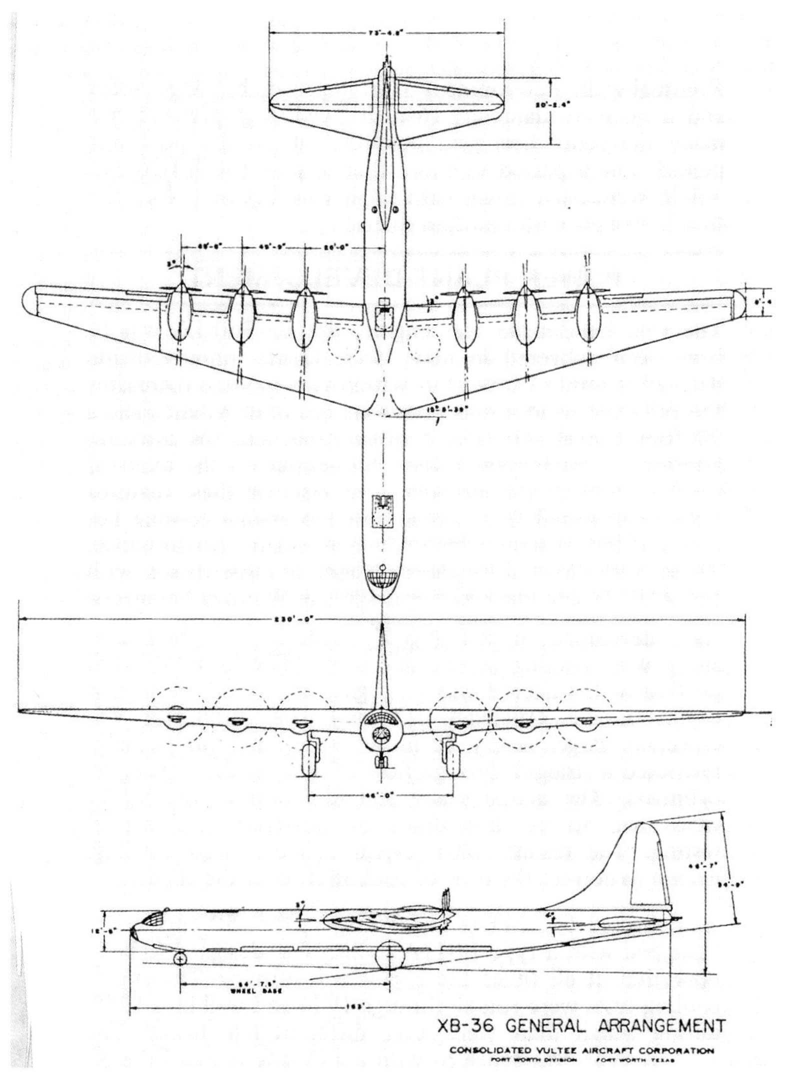

XB-36 3-view drawing of the plane's configuration. (Convair)

Nearing completion in April 1946, the XB-36 has had its engines and propellers installed. Notice the American flag that hangs on one of the huge sliding doors of the Experimental Building. (Convair)

Two views of the finished XB-36 as it prepares for a series of ground and taxi tests prior to its first flight. Notice the tall single tail that replaced the original twin tail shown in early design studies. Summer 1946. (ACME)

The XB-36 lifts off on its maiden flight, August 8, 1946, in front of Convair Fort Worth. Beryl A. Erickson and Harold "Gus" Green are at the controls. Notice the giant single tire main landing gear that restricted the XB-36 to only three specially strengthened airfields in the entire country—namely Fort Worth, Eglin Proving Ground in Florida, and Fairfield-Suisun (later Travis AFB) in California. (Convair)

The right landing gear side brace strut broke loose on take off during the 16th test flight of the XB-36 on March 26, 1947. An emergency was declared, and the #4 engine was feathered. The fate of the entire B-36 program was in jeopardy until Convair pilots Erickson and Green landed the big bomber safely. (Convair)

Two views of the XB-36 in flight. The clean, aerodynamic lines of the aircraft can be appreciated in these photographs. (Convair)

Convair employees and Air Force personnel inspect the damage to the right landing gear. Erickson had brought the XB-36 in and "bicycled" down the runway on the damaged gear. When the plane ran off the runway, the 110" diameter tire sank only 9 inches into the soft dirt. Broken side brace strut clearly visible. (Convair)

The XB-36 revving up its engines in 1950 with a new track-type experimental landing gear. The gear was not intended for production models, but was being tested to prove the feasibility of a track-type gear on a plane as large as a B-36. It did make a loud screeching noise when it took off for the first and only time on March 26th. (Convair)

Mockup of new B-36 nose design with raised canopy for better visibility, June 1945. Notice dummy 20mm cannon frontal gun installation. An earlier nose armament proposal in 1944 included barbette gun turrets mounted on each side of the nose right below the cockpit. (Convair)

Overhead view of new bubble-type glass canopy that covered the pilot, co-pilot, and flight engineer, June 1945. The redesigned layout was more efficient than in the XB-36. Notice real metal seats in basically an all-wood mock-up. (Convair)

The YB-36, 42-13571 was the production-standard B-36. It is seen here with its tail being tilted downward in order to clear the factory doors. Notice the propellers have not yet been installed and the new nose gun position is covered. (Convair)

The YB-36, second prototype, parked in the north yard at Convair Fort Worth in June 1947. It has the same single wheel main landing gear as the XB 36. It would not fly until December 4th. The new nose design with raised canopy would become standard on all subsequent production models. With improved equipment such as GE BH-2 superchargers, the YB-36 soon outperformed the XB-36, surpassing the XB-36's highest altitude by reaching 40,000 ft. on only its third flight. (Convair)

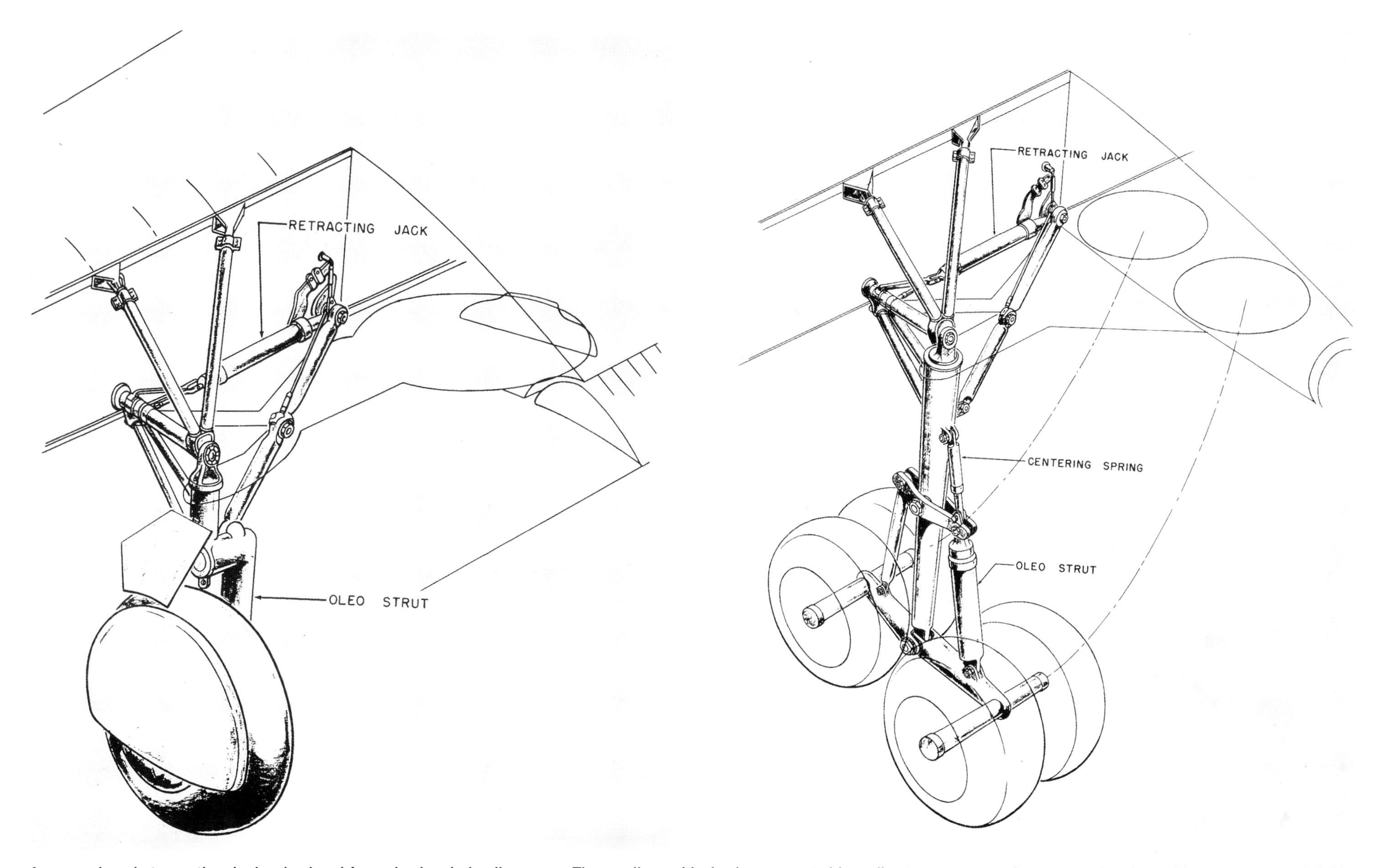

A comparison between the single wheel and four wheel main landing gears. The smaller multi-wheel gear exerted less direct pressure on the runway, thereby making many more airfields accessible to the heavy bomber. (Convair)

B-36As on the production line in Fort Worth during 1947. The first B-36A in the foreground is 44-92004, which was reserved as the static test example. The second aircraft down the Convair line is 44-92005. In all 22 B-36As would be completed by December 1948. (Convair)

B-36A, 44-92004, flew to Ohio on August 30, 1947. Col. Tom P. Gerrity was the pilot, Beryl Erickson was an observer. The plane carried only equipment necessary for the one time flight. Landing at Wright Field after a 4 hour 40 minute flight, the second B-36 ever to fly was gradually torn apart during static structural tests in a specially equipped hangar. (Convair)

B-36A, 44-92009, on a Convair shakedown flight in early 1948. Like the YB-36 and the other 21 B-36As, it carried no armament. Notice the early football-shaped radio compass antenna under the nose and the "buzz number" BM-009. (Convair)

This underside view of B-36A, 44-92006, in flight shows off the immense size of the bomber with its 230 ft wing. (Peter Bowers/David Menard)

B-36A, 44-92022, cruises over the countryside on a training flight. BM-022 was used for training and crew familiarization, as were all the B-36A models. (USAF)

BM-005 getting ready to land with its gear extended. Plane was assigned to Air Materiel Command (AMC) at Wright Patterson AFB, Ohio, for evaluation. (Author's collection)

Two views of B-36A, 44-92015, the first B-36 delivered to the Strategic Air Command on June 26, 1948. Notice the triangle symbol on the tail (8th Air Force) and the name painted on the nose, "City of Fort Worth." Location of airfield is unknown, but appears not to be Fort Worth (San Diego Aerospace Museum)

A youngster having specifications of the B-36A explained to him by a crewman at the 1948 International Air Exposition, which celebrated the opening of New York's Idlewild (later John F. Kennedy) International Airport. Sliding panels of the forward bomb bay partially obsure the buzz number BM 015, the "City of Fort Worth." (USAF)

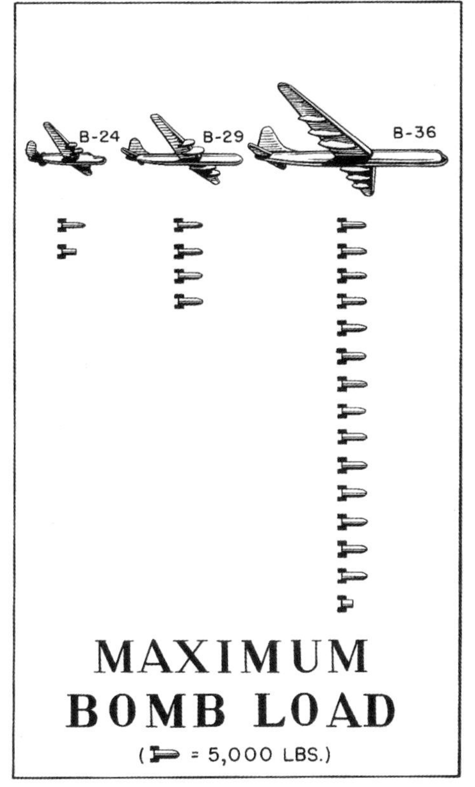

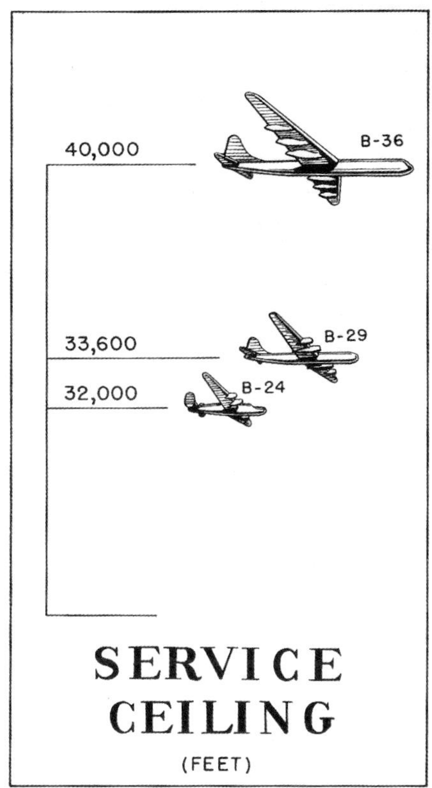

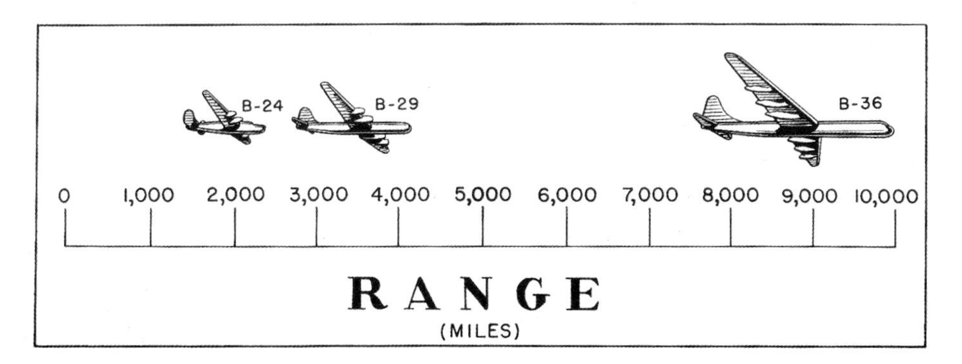

Performance chart comparing World War II heavy bombers to the Air Force's newest intercontinental bomber, the B-36. (Convair)

Six Pratt & Whitney R-4360-25 Wasp Majors powered the B-36A. Each air-cooled engine developed 3,000 hp. (Author's collection)

Left wing engines, #1, #2 and #3 showing the ring-shaped airplugs. Exhaust gases from the engine were vented from the bottom of the nacelle. The #1 engine in the foreground has its airplug partially open. The number of diamonds visible to the flight engineer indicated the airplug's position. (RKO/ Walter Jefferies)

An airman ponders the size of the 19 ft. propellers on #4, #5 and #6 engines on the right wing of a B-36A at Carswell AFB. Notice the C-54 transports and World War II era B-29s in the distance. (Author's collection)

Mechanics working on a B-36 powerplant fitted with the more efficient square-tipped propellers. (Author's collection)

Engine change is underway on 44-92060. With 28 cylinders and 56 spark plugs, keeping each B-36 engine running was a demanding task for Air Force mechanics. The Pratt & Whitney R-4360 Wasp Major radial engine was the ultimate development of the piston engine—future SAC bombers would be powered by jets. (USAF)

Unveiling of the new Carswell Air Force Base sign on a obviously windy winter day, January 30, 1948. Formerly known as Fort Worth Army Air Field, the base was renamed in honor of Congressional Medal of Honor winner, Maj. Horace S. Carswell, Jr., of Fort Worth. Carswell was a B-24 pilot shot down while attacking a Japanese cruiser and destroyer. Maj. Gen. Roger M. Ramey, 8th Air Force commander, salutes Mrs. Carswell as he walks to greet her. (USAF)

Three new B-36Bs for the 7th Bomb Group on the flightline at Carswell. The B-36B model had water-injected 3,500 horsepower Wasp Majors and was the first fully combat equipped model. The aircraft in the foreground is 44-92040. (USAF)

The 30th airplane off the Convair production line, B-36B, 44-92033, is piloted by AMC flight acceptance officer Maj. Stephen P. Dillon on a test flight. All B-36Bs had defensive armament, which included sixteen 20mm cannons in eight turrets. The turrets were retractable, except for the nose and tail. (Convair)

The 7th Bomb Group ramp at Carswell AFB, TX, shows a number of B-36As and red-tailed B-36s in 1949. B-36As BM-013 and BM-025 appear to have each other's rudder. (USAF)

The 11th Bomb Group flightline, also at Carswell in spring 1949. Two of the aircraft have red tails, indicating they are part of the 18 plane Gem program that involved cold weather testing in Alaska. The B-36B in the foreground, 44-92032, would later crash at Fairchild AFB in 1954 when assigned to the 92nd Bomb Wing. (USAF)

Below Left: Bomb bay of the B-36 was cavernous compared to World War II bombers. Capacity was 72,000 lbs. of bombs, or two 43,000 lb. "Grand Slam" bombs. Two airmen are seen here working on the bomb racks in the forward bomb bay. Just behind them, another airman passes under the huge main wing spar box. (USAF)

Right: B-36B, 44-92077, is framed by the wing and fuselage of another 7th Bomb Wing B-36B. The triangle symbol on the tail meant the plane was assigned to the 8th Air Force. Later a letter J would be added to indicate 7th Bomb Wing, or a letter U for 11th Bomb Wing. (RKO/Walter Jefferies)

Amusing illustrations from B-36A and B-36B flight manuals. Later B-36H and B-36J manuals were more reserved with less cartoon-type drawings. (USAF/Scott Deaver)

B-36Bs from Carswell flying formation over the Texas countryside. B-36s in formation were an impressive but rare sight. Combat missions were based on a one plane-one target concept. Top speed of the B-36B was 381 mph (USAF/Don Bishop)

The Convair Fort Worth plant as seen from the air in 1949. Carswell AFB is just across the field in the distance. The sole XC-99, cargo and transport version of the B-36, can be pinpointed just under the wingtip of the photo plane in the picture, along with six B-36s in the north yard area. (USAF)

One of five Carswell B-36s flies low over the Capitol during the Inauguration of President Harry S. Truman on January 20, 1949. (USAF)

President Truman shakes hands with a crewmember of a Carswell B-36 from the 26th Bomb Squadron, 11th Bomb Group, that participated in an eleven plane aerial demonstration at Andrews AFB on February 15, 1949. The demonstration was for the benefit of the President and members of the U.S. Senate at their request. (USAF)

President Truman waving to photographers from an open entry hatch on a B-36 Peacemaker put on display at Andrews AFB for his personal inspection. The name Peacemaker was never adopted officially by the Air Force—it was judged the winner of 60 entries in a contest sponsored by Convair among its employees in spring 1949. (USAF)

An early study of the B-36B airplane which proposed increasing its speed by using eight XT35 Curtiss-Wright gas turbines placed in four tandem nacelles. Top speed would supposedly have been 448 mph. Convair offered to install the gas turbines on one test B-36 in February 1947, but nothing came of the proposal. (Convair/ Bill Plumlee)

Model of the proposed YB-36C which would have used variable discharge turbine (VDT) engines based on the standard R-4360. Top speed was estimated to be 410 mph. However, it would have been necessary to redesign the B-36 from a pusher-type configuration to a tractor-type version. Problems with adapting the VDT engine to the B-36's wing eventually caused the cancellation of the program in spring 1948. (Jones collection)

Prototype B-36D, 44-92057, rotates on take off with its six reciprocating engines augmented by four Allison J35-A-I9 jets, same as those used on the XB-47. It flew first on March 26, 1949, with J35s since production General Electric J47-E-19s were not yet available. Notice lack of bracing struts on the jet pods. (Convair)

Front view of the B-36Ds new jet nacelle with iris blades closed. Unlike the J-47s on the B-47 bomber, B-36 jet engines would not be used continually in flight, and consequently were shut down when not needed. The iris blades controlled air intake and prevented windmilling when not in operation. The boost from jets was mainly used for take off, climbing, and speeding over a target. Notice the side brace strut, added after vibration problems were encountered. (RKO/Walter Jefferies)

Close-up of one of the prototype B-36D's jet pods. Each J47 jet developed 5,200 pounds of static thrust. 057 flew with production J47s installed on July 11, 1949. The technician is plugging in a ground generator. Notice the taxi light, same as those on the B-47 pods. (Convair)

A competitor of the B-36, the Northrop eight jet YB-49, makes its maiden flight at Hawthorne, California, on October 21, 1947. At one time, production of B-49s was planned at Convair Fort Worth, but later canceled in favor of ordering more B-36s. (Edwards AFB Office of History)

Several of America's newest bombers were gathered together for a congressional viewing at Andrews AFB during the 1948/1949 debate on which aircraft should be given the green light for further production. This side view of the YB-49 shows off its sleek lines and positions it ahead of the B-36. However, in reality, it fell behind the B-36 in the range category and had unresolved stability problems. The Northrop Flying Wing concept was later to be resurrected in the 1980s with the computer-assisted B-2 Stealth Bomber. (Edwards AFB Office of History)

Artist rendering of the *USS United States*, CVA-58. The 65,000 ton ship was to be the first of four in a new class of flush-deck supercarriers. When it was abruptly canceled by Secretary of Defense Johnson in April 1949 in order to purchase more B-36 bombers, the ensuing controversy led to a series of congressional hearings in August and October. However, the charges of impropriety in procurement of the B-36, and favoritism toward the Air Force's stated role in providing the nation's sole strategic bombing force were dismissed, and both Convair and the Air Force were cleared. The charges by the Navy that the B-36 was a "sitting duck" and could not adequately defend itself against enemy jet fighters was explored but not resolved. It was decided to continue with the B-36 program since it was still the best bomber available at the time. (National Archives)

Artist rendering of a RB-36 being "attacked" by Navy Banshee jets. The Navy claimed it could easily shoot down the lumbering B-36. A test duel between the two was proposed but never conducted, canceled by the Secretary of Defense for national security reasons. (D. Sherwin)

B-36Bs at Convair San Diego for the "B to D" conversion program. Fifty-four B-36Bs were brought up to B-36D standards in San Diego, and five were converted in the Fort Worth plant. The program started in April 1950 and ended in February 1952. (Convair)

Major modification work on B-36s at the San Diego Lindbergh Field plant included adding jet pods, new quick-action bomb bay doors, metal control surfaces, and interior equipment upgrading. Aircraft on the right is B-36B, 44 92043, which had dropped two 42,000 pound "Grand Slam" conventional bombs in a January 1949 demonstration. (Convair)

Converting B-36Bs to jet-augmented D models was done in an outdoor work area. Four red-tailed B-36s, formerly in the Gem program, can be seen here at the Lindbergh Field facility. The coding sequence, FW-SD, indicates the airframe's position in the original Fort Worth production sequence and its position on the San Diego line. (ACME)

New quick-action bomb bay doors were installed on the B-36Bs and the B/RB-36Ds then being built in Fort Worth. The two section hinged door panels opened and closed hydraulically in just two seconds. All subsequent B-36 models would feature these doors that replaced the former sliding panels previously used on the two prototypes, B-36As, and B-36Bs. (USAF)

Night shift at Convair San Diego in 1950. Planes were too large to be brought inside the old factory building, so "B to D" modification work was done mostly outside in the adjacent yard area. Security was of concern at the time since the Korean War had just broken out in June, and raged during the entire conversion program. Espionage or sabotage by Communists was considered a real possiblity. (Convair)

B-36B, 44-92057, the prototype B-36D, is shown here on a test flight over Fort Worth. Notice the production sequence number 54, clearly visible on the nose. (Convair)

RB-36D reconnaissance version of the B-36. It is easy to see in this photograph why one of the nicknames of the B-36 was "magnesium monster." The dull sections of the fuselage, mostly the bomb bay area, are lighter magnesium metal to decrease weight and thereby increase range. Notice the three ECM antennas under the bomb bay area. Later these will be moved further aft in order for the plane to carry bombs in the rear bays. (Convair)

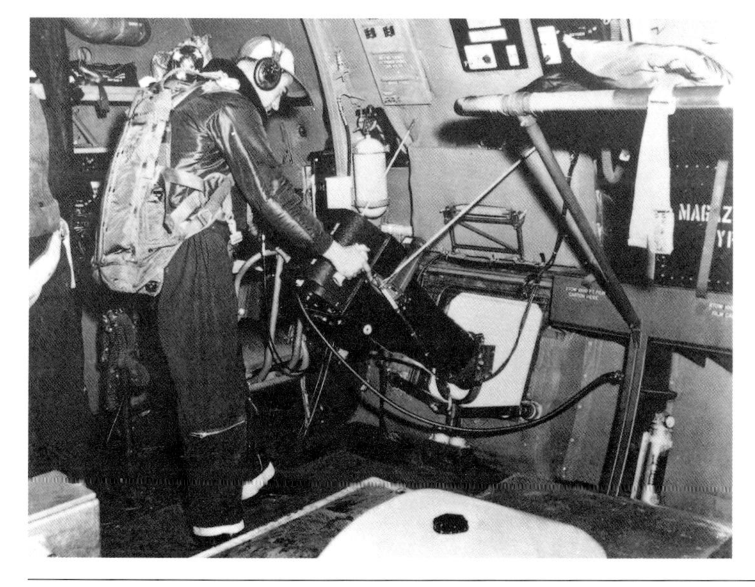

Left: A RB-36 photographer aims his oblique camera from the forward cabin compartment. (USAF)

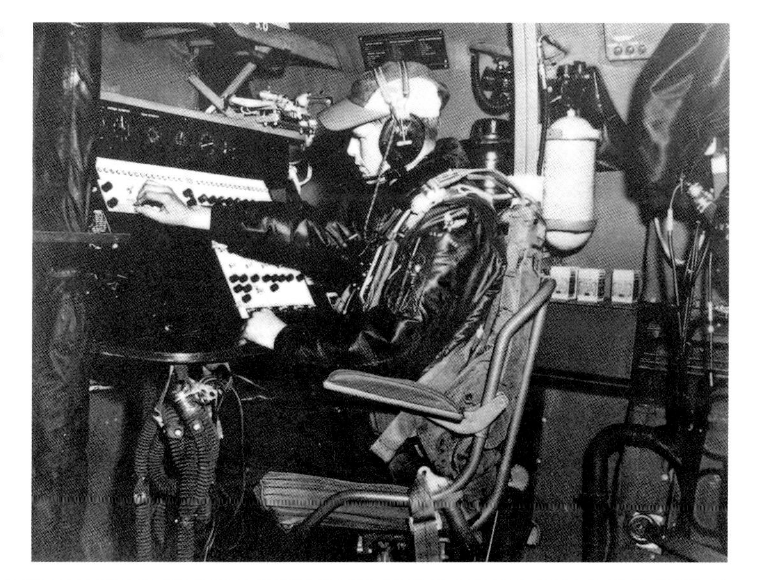

Right: Photo technician at his console in a RB-36. Film could be developed in flight if necessary, using a small on-board darkroom. (USAF)

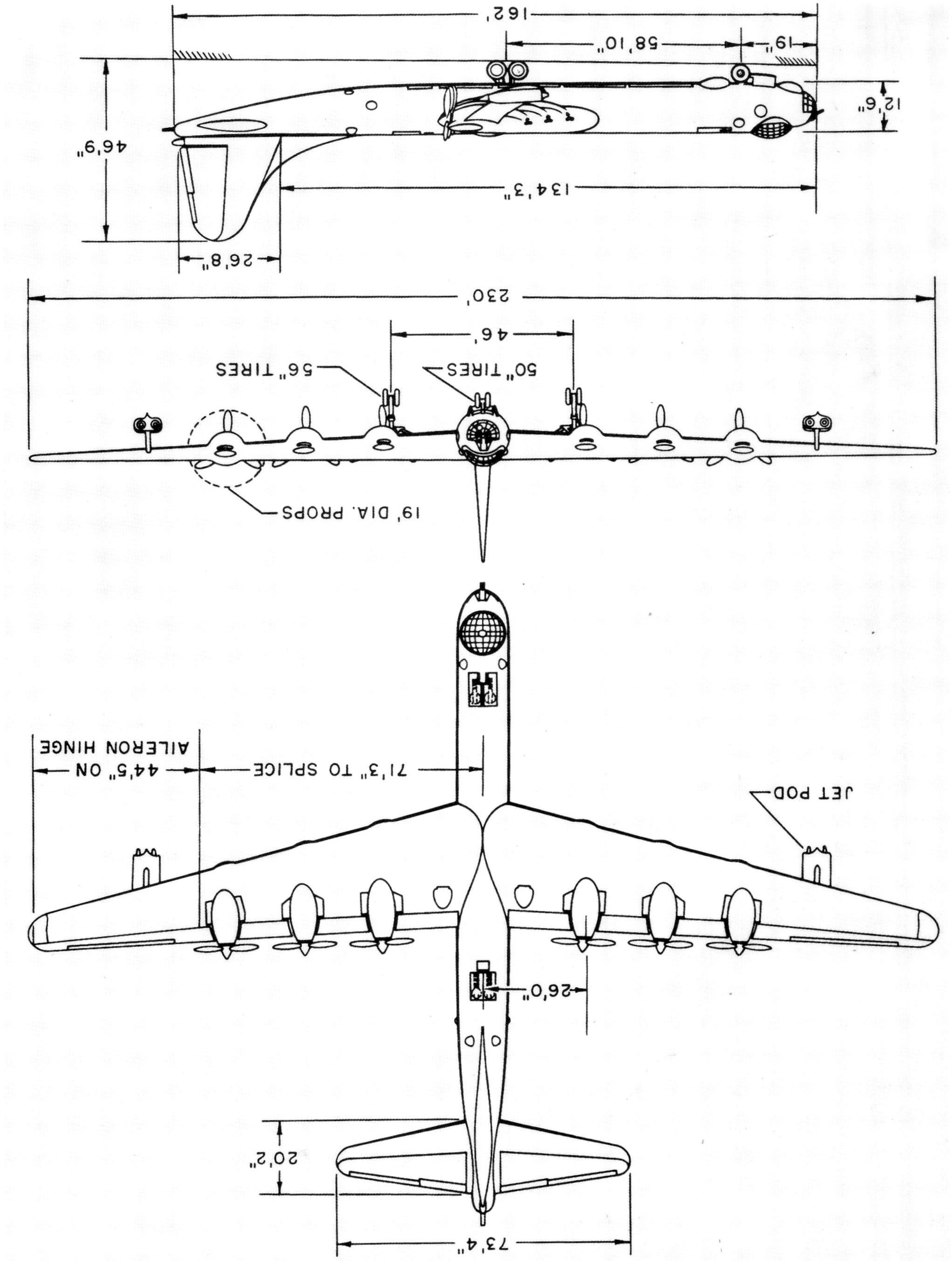

3-view plan of Convair B-36D. (San Diego Aerospace Museum)

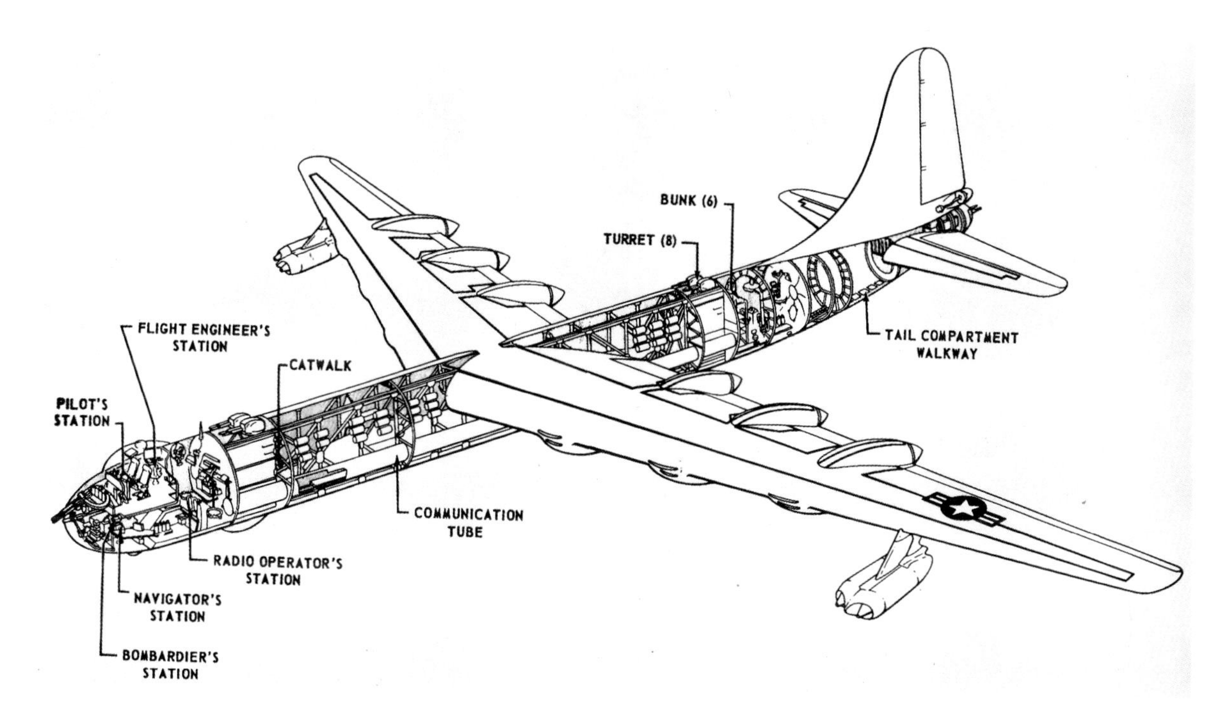

General arrangement drawing of the B-36D. The 85 ft. communication tube connected the forward and rear pressurized crew compartments. To travel through the tunnel a crewman had to lie on his back on a small cart and pull himself along by means of an overhead rope. Most crewmen did not like to use the tube, since it was dark, eerie, and noisy. It also could be dangerous if decompression ever occurred. Small windows in the tube permitted inspection of the bomb bays at pressurized altitudes. Otherwise, inspection could be done by using a narrow catwalk running along the opposite side. (Convair)

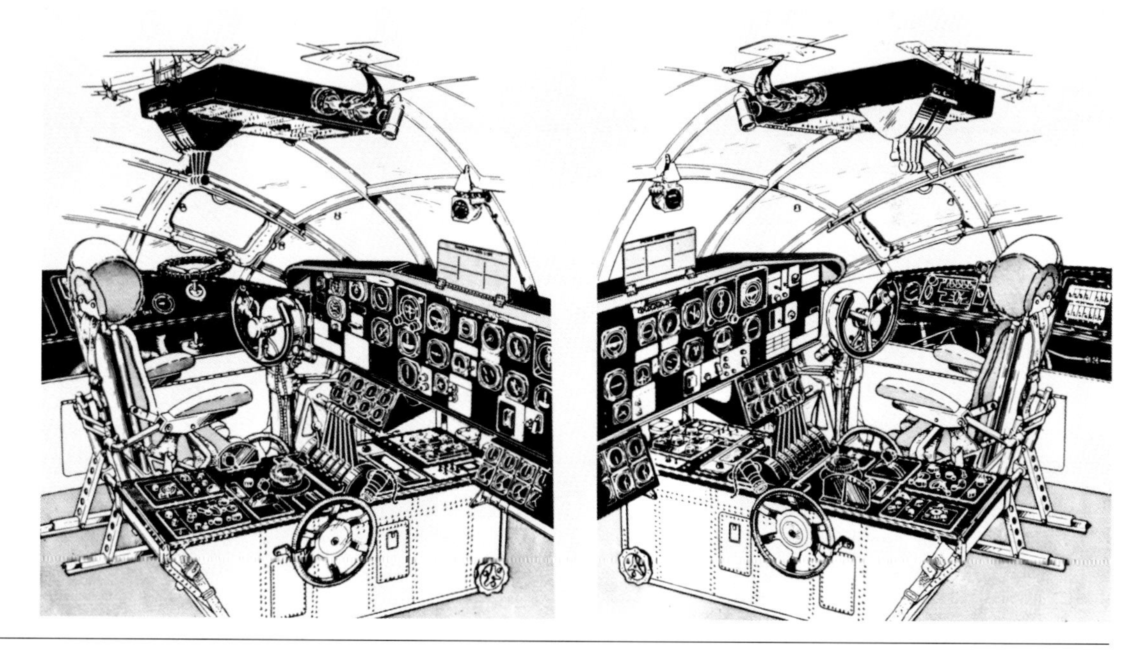

Pilot's and co pilot's stations on the B-36D flight deck. Notice the controls for the auxiliary jets located above on the canopy, and the wheel to the left of the pilot, for nose wheel steering. (Convair)

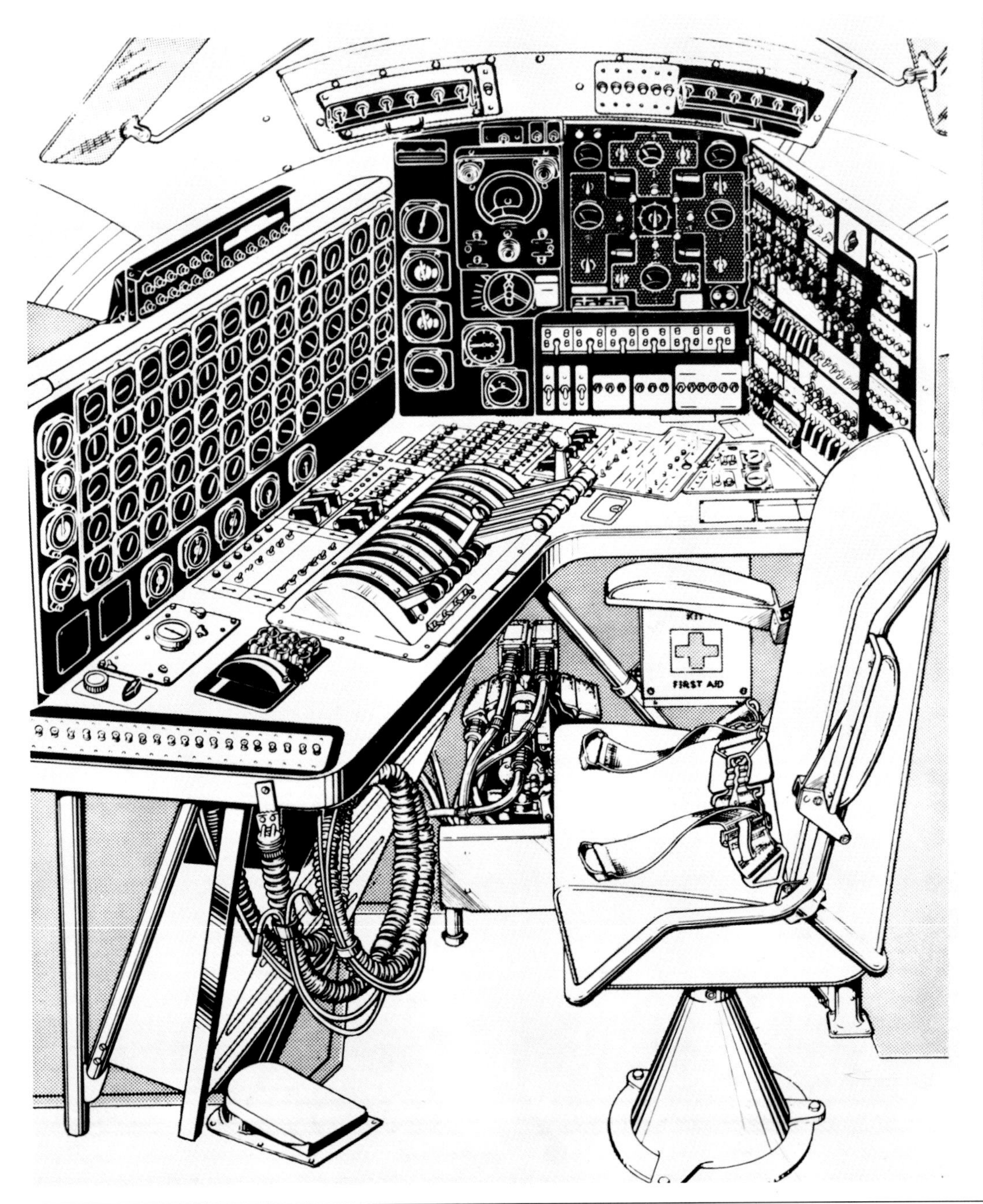

Two views of a RB-36H at an open house at San Francisco International Airport.

It displays its sixteen 20mm cannons to the public: the two-gun nose turret, the two upper forward retractable turrets each with two guns, and the upper and lower aft retractable turrets each with two guns. All six retractable turrets carried 600 rounds each. The tail turret, like the nose turret, had two guns and handled 400 rounds. This array of turrets made the B-36 the most heavily-armed bomber in history, but the effectiveness of this defensive system was never actually tested in combat. (Larkins collection)

Left: B-36D flight engineer's station. Pilot's throttle controls are duplicated for the flight engineer. Fuel mixture levers are to the left of the longer throttle controls, while those needed for the electrical system are to the right. (Convair)

Close-up of a B-36's nose showing stowed 20mm cannons, nose gun sight, UHF antenna, and oval window for optical bombing which was later made obsolete by K-system pariscopic sight and radar. (RKO/ Walter Jeffries)

An airman checks the guns in the forward upper turrets. Notice that the access panels on the turrets have been removed for servicing. (Author's collection)

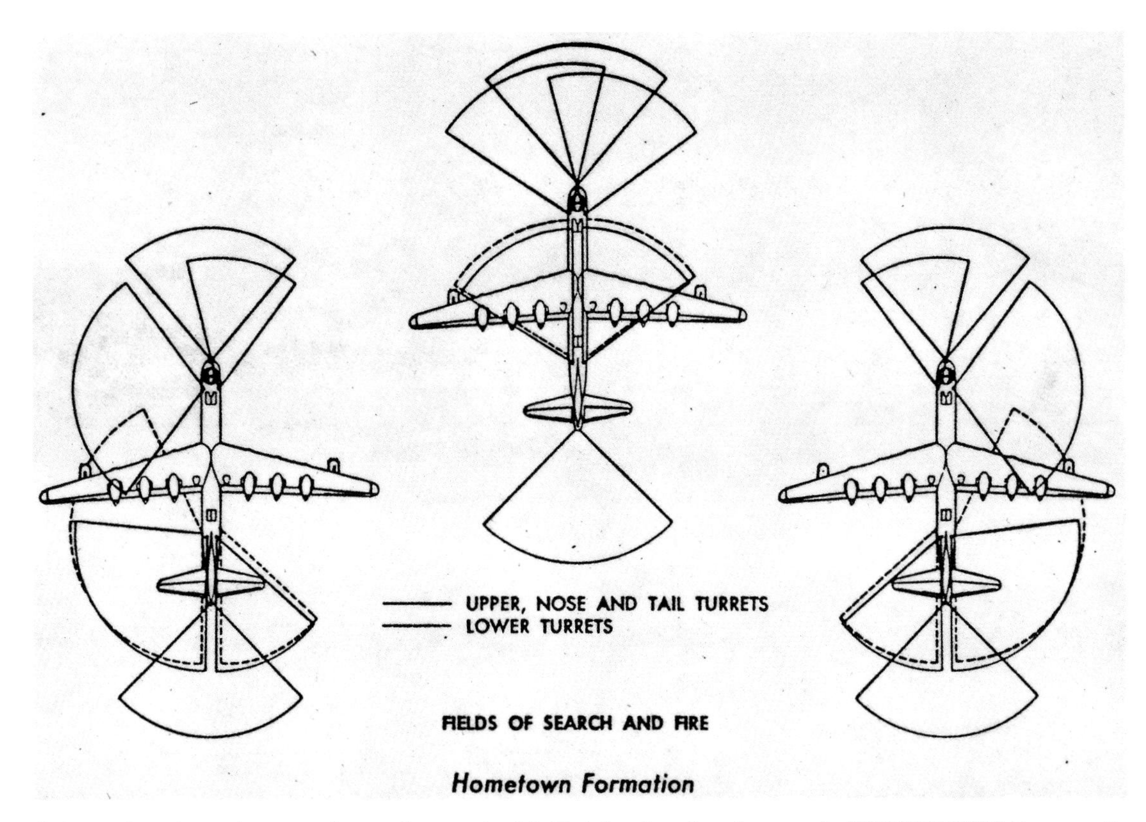

A three-plane formation, or cell, was the standard B-36 defensive plan. It was called "HOMETOWN." However, B-36s rarely flew in formation, since each plane was usually assigned its own target with a specific mission plan. (USAF)

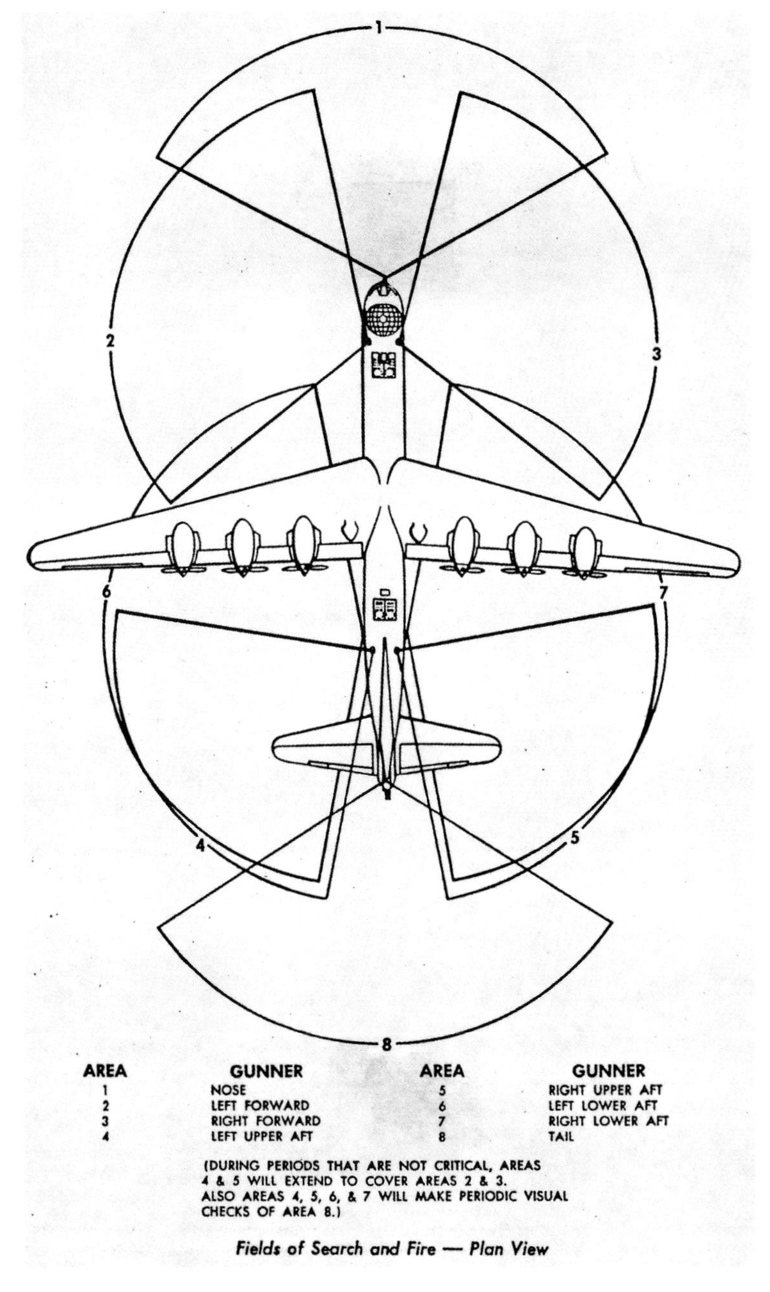

Plan view of fields of search and fire for the B-36. (USAF)

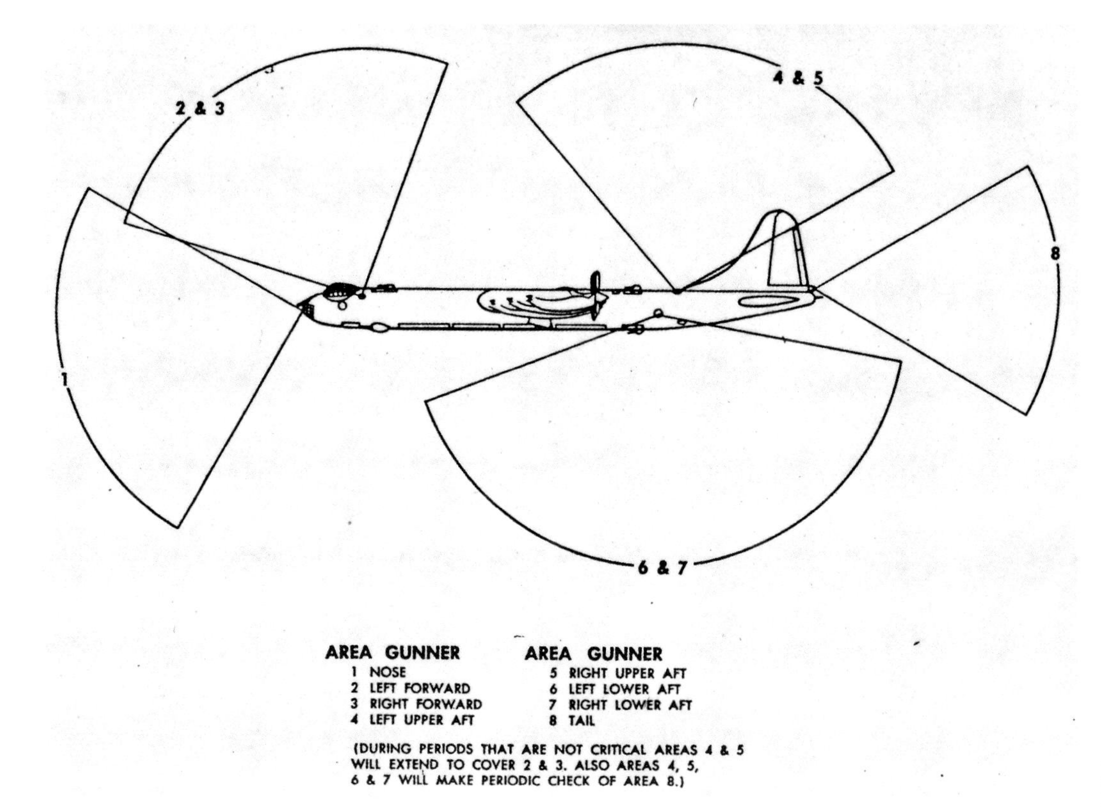

Side view of the B-36's fields of search and fire. (USAF)

Five hundred pound bombs loaded on racks in a bomb bay of a B-36. Photo was supposedly taken before a bombing demonstration held at Eglin AFB Proving Ground in Florida. (7th Bomb Wing B-36 Association)

RB-36E, 44-92020, from the 5th Strategic Reconnaissance Wing at Travis AFB, California. All 21 B-36As and the YB-36 prototype were converted to RB-36E reconnaissance aircraft, which were basically identical to the RB-36D. Notice the circle X symbol on the tail, which signifies it as assigned to the 5th SRW and 15th Air Force. (Bodie collection)

Another view of RB-36E, 44-92020, on the ground several years later. Notice the circle X tail code has disappeared (tail codes were dropped in 1953), along with the 15th Air Force emblem. Also the "United States Air Force" lettering on the fuselage has been replaced by much large letters. (Merle Olmsted)

Seen here a 1957 flight to Wright Patterson AFB, Ohio, carrying a B-58 airframe attached to its bomb bay, is B-36F, 49-2677. The propellers were removed on engines #3 and #4 to acccomodate the B-58, which was minus its four J79 jet engines and other equipment. It was scheduled to be torn apart in static structural tests like the B-36 had experienced ten years earlier. B-36Fs had more powerful 3,800 hp Wasp Major engines with a top speed of 417 mph, with a service ceiling of 44,000 ft. (USAF/Don Bishop)

The B/RB-36H models, also with improved 3,800 hp engines, were the most produced of the B-36 series. In all, 156 examples were built by Convair. Here B-36H, 50-1092, is exhibited at a Detroit air show on September 1, 1952. The triangle U symbol on the tail indicates the plane is from the 11th Bomb Wing, 8th Air Force. (AAHS/William Steeneck)

Under guard at Ellsworth AFB, South Dakota, in May 1956, is RB-36H, 51-13717. The plane is from the 28th Strategic Reconnaissance Wing. Notice the three ECM antennas are located as far back as possible to allow use of the aft bomb bay when the RB-36s' primary role changed to bombing. (Merle Olmsted)

B-36H, 52-1359, with its two sets of bomb bay doors open, sits on an unidentified flightline. The underside of the fuselage is painted with special, anti-flash white paint to help protect the plane against the effects of a nuclear blast. (Merle Olmsted)

A 7th Bomb Wing B-36B flying low over the Gulf of Mexico. Notice the shadows on the water of two other bombers in the formation. (7th Bomb Wing B-36 Association)

SAC operations became global in nature. One of six B-36s from the 7th Bomb Wing is shown here in the snow at Limestone (later Loring) AFB in January 1951. This particular aircraft is B-36D, 49-2652. All of the B-36s staged through the Maine base for the overwater flight to the United Kingdom. This airplane earlier sported phoney markings in 1950 when it was going to be featured in the aborted RKO motion picture, "High Frontier." (USAF)

Another B-36 making the same overseas flight originating from Carswell AFB in January 1951. Dubbed "Operation UK," B-36D, 49-2658 flew with five other B-36s from the 7th Bomb Wing to RAF Lakenheath. It was the first time a B-36 was ever seen in Europe. Russian diplomats in London were impressed with its massive size. (USAF)

B-36H, 52-1366, being refueled on the snow-covered flightline at Loring AFB. Notice the lethal AN/APG-32 tail gun system with the twin radomes enclosed in a single cover. (USAF)

Huge hangar built at Rapid City Air Force Base (later Ellsworth AFB) in South Dakota. It could easily accomodate two B-36s with room to spare. (Author's collection)

RB-36H, 49-2688, out of Rapid City AFB. Name of the base was changed to Ellsworth AFB in honor of base commander Brig. Gen. Richard E. Ellsworth, who was killed with 22 others in a 1953 crash of a RB-36 in Newfoundland. (Author's collection)

"GCA" ALL THE WAY. A little "humor" from the photo lab at Ellsworth AFB. (USAF/Robert L. Bartlett)

B-36D, 44-92065, originally a B-36B. Markings show it is assigned to the 92nd Bomb Wing at Fairchild AFB, Washington. The circle W indicates 15th Air Force, and the W, 92nd BW. It also carries an "Alley Oop" cartoon figure on the nose belonging to the 326th Bomb Squadron. (David Menard)

Carrying the markings of the 7th Bomb Wing, 8th Air Force in September 1950, this B-36D experienced two severe occurences during its career. It was first damaged in the tornado that ravaged the Carswell flightline on September 1, 1952. Later, when assigned to the 95th Bomb Wing, it crashed at El Paso, Texas, in 1954. One crewman was killed, several injured. (USAF)

Fifteen man crew of a B-36 checking each other's equipment during a pre-flight inspection prior to a 20 to 40 hour mission. (USAF)

All the comforts of home! Preparing breakfast aboard a B-36, which had a small stove and oven. Later, during "featherweighting," the galley was tossed out. (USAF)

The 23 man crew of a 72nd Bomb Wing RB-36 prepare for inspection before boarding the aircraft. The 72nd was based at Ramey AFB, Puerto Rico, the only B-36 base located outside the continental United States. (USAF/Jim Ballard)

An aerial view of Ramey AFB in the mid-fifties. Two RB-36s can be seen parked near the base crash and rescue services building. Ramey was known as the "country club" of B-36 bases due to its tropical location. (Richard "Pogo" Graf)

Head on view of B-36F, #1064, revealing a second B-36 from the 6th Bomb Wing, Walker AFB, New Mexico, parked right behind it. Notice these are featherweighted B-36Fs, as is evidenced, in part, by the lack of nose turrets. (Thomas Gannon)

Above: Dramatic shot of a RB-36 from the 99th Strategic Reconnaissance Wing in flight to Yokota AFB, Japan, during a deployment in 1956. (James W. Church)

Creating a cloud of dust, another RB-36 from Fairchild AFB backed into its parking place after arrival at Yokota AFB. (James W. Church)

"Six Turnin' and Four Burnin'" personified. B-36J, #2225, assigned to the 11th Bomb Wing, at altitude over the Mediterranean during the Lebanon crisis in 1956. This particular plane was also the winner of the Fairchild Trophy for bombing accuracy in 1956. (Reginald M. Beuttel, Jr.)

A flight of RB-36s from the 72nd Bomb Wing, returning after a 1955 deployment to Turkey. The aircraft returned singularly and rendezvoused near Puerto Rico to form up and return in formation. Notice the blackened horizontal stabilizers...oil from many hours of flying. (Robert M. Cameron)

Remains of B-36B, 44-92079, which crashed into Lake Worth during a night take off on September 15, 1949. Five of the 13 crewmen were killed. It was the first crash of a B-36 resulting in fatalities. Notice the Convair factory in the distance. (7th Bomb Wing B-36 Association)

Two B-36s of the 7th Bomb Wing tossed into each other during the tornado that struck Carswell AFB on September 1, 1952. A total of 106 planes were damaged, including some across the field at the Convair plant. All B-36s, except one beyond repair, were back in service by May 11, 1953. (USAF)

Two views of the damage done to a B-36H by the severe storm that hit Carswell in 1952. The 7th Bomb Wing plane had its cockpit and forward fuselage damaged, but rather than make repairs (many of the other B-36s damaged in the storm were fixed or returned to Convair for rebuilding), 51-5712 became the airframe for the experimental NB-36H nuclear reactor test plane. The reason for the decision was simple: why pull a B-36 off the production line to install the special nose crew capsule the NB-36H required, when 5712 already needed a new nose? (7th Bomb Wing B-36 Association)

Not really an accident, but more of an embarassment, RB-36H, 51-13730, was tilted back on its tail by a gust of wind while on display at Chanute AFB, Illinois, in May 1957. Today, 730, after relocation in 1991 to the Castle Air Museum in central California, is back on exhibit surrounded by aircraft contemporary to its era, such as a B-47 and B-52. (George E. Mayer)

Aerial view of the Fort Worth plant—estimated to be 1953 or 1954. Originally built by the U.S. government, the mile-long factory first produced B-24s, then B-32s for the war effort. B-36 production was to last from 1947 through 1954. Over the years, the plant has seen several company names, including Consolidated, Convair, General Dynamics Convair Division, Lockheed, and most recently, Lockheed Martin. The B-36 was the largest airplane ever built at the facility, and the huge buildings have produced a number of smaller size military planes, such as the B-58, F-111, and F-16 built in later years to present. (San Diego Aerospace Museum)

Being towed at Lindbergh Field at Convair San Diego, RB-36D, 49-2695, has completed maintenance in the SAM-SAC program. Sequence numbers on the nose indicate it is the 119th plane built at Fort Worth and the 78th to undergo SAM-SAC maintenance and equipment updating. This particular plane would later be modified into one of the ten GRB-36s assigned to the 99th Strategic Reconnaissance Wing. (Convair)

Dramatic mass flyover of Carswell B-36s and military parade sequence from Convair-produced documentary film, "Target: Peace." It premiered at the Spreckels Theater in San Diego on October 19, 1949. The film showed the American public the giant B-36 bomber for the first time and explained its role in helping keep the nation at peace. (San Diego Aerospace Museum)

L
92652
UNITED STATES AIR FORCE
001

UNITED STATES AIR FORCE
001
92652
L

"Star," of the Paramount film, "Strategic Air Command," B-36H, 51-5734. Plane was from the 26th Bomb Squadron, 11th Bomb Wing. Actually two other B-36s were also painted with number 5734 (for backup) at one time. (USAF)

5734 (in foreground) on the flightline at Carswell AFB, Texas, where most of the B-36 sequences were filmed for the 1955 motion picture, "Strategic Air Command." A 7th Bomb Wing B-36 is being refueled in the background. (Author's collection)

Opposite: Two views of B-36D, 49-2652, which was going to be used in the RKO film, "High Frontier"—starring Richard Widmark, Fred MacMurray, Claude Rains and Ann Blyth. Most of the markings were false, such as the triangle L on the tail (There was no bomb wing identified with the letter L). Although the serial number on the tail fin is accurate, the 001 buzz number does not match. The 18th Air Force emblem on the tail is fictional, and finally, the unit badge on the nose is fake, too. Russian spies must have been confused! (RKO/Walter Jefferies)

Actor Jimmy Stewart watches Harry Morgan monitor the flight engineer's instruments, while Barry Sullivan, as co-pilot, reads the check-off list prior to take off.

B-36J, 52-2827, the last B-36 off the assembly line on August 14, 1954. Notice the factory sequence number has been blacked out for security reasons. All B-36s had to be tilted nose upward in this manner in order for the tail to clear the top of the doorway. Delivery of 2827 completed SAC's B-36 fleet and ended an era. (General Dynamics Convair)

Almost five years later, 2827 was flown to Amon Carter Field in Fort Worth from Biggs AFB for retirement ceremonies on February 12, 1959. It was the last SAC B-36 mission ever flown. (General Dynamics Convair)

Placed as a memorial to the personnel who built, maintained, and flew the B-36 bomber, 2827 was put on display at Greater Southwest Airport, now a business park south of Dallas/Fort Worth International Airport. Today, the plane is stored in several hangars, disassembled in sections, and awaiting construction of a new aviation museum at Alliance Airport. (General Dynamics Convair)

Having done their duty in SAC, most of the B-36 fleet awaits its fate in the hot desert sun at Davis Monthan AFB in Arizona. The stripped carcasses will be systematically chopped apart and smeltered down to aluminum ingots. By early 1961 they all were gone. (USAF)

Flying into the wild blue yonder for a final flight, a B-36J heads north and eastward from Davis Monthan to the Air Force Museum in Dayton, Ohio. The flight took place on April 30, 1959, and was the very last time a B-36 was ever seen or heard in the skies over America and the rest of the World. (Author's collection)

Wasp Major engine being removed from a B-36 undergoing reclamation in April 1957 at Davis Monthan. The R-4360s could still be used on other Air Force aircraft, such as the C-124 Globemaster. (MASDC/Scott Deaver)

Model of proposed commercial transport version of the B-36, called Convair Model 37, like the XC-99, it would use the same wing, engines, and landing gear of the bomber, but would have had a double-deck fuselage and taller tail. Pan Am originally ordered 15 airliners in 1945, but took options on only three, finally canceling altogether. The plane was designed to carry 204 passengers and 8 tons of cargo between London and New York in only nine hours. (Consolidated Aircraft Corporation)

The XC-99 cargo/transport under construction at Convair San Diego's plant in 1947. The double-deck fuselage was built in San Diego, with the wing, engines, and landing gear shipped from Fort Worth. The engines installed were the same 3,000 hp Wasp Majors as on the XB-36. As a troop carrier, the XC-99 was to transport 400 fully combat equipped soldiers. As a cargo plane, it was designed to carry 100,000 pounds of cargo on its two interior decks. (San Diego Aerospace Museum)

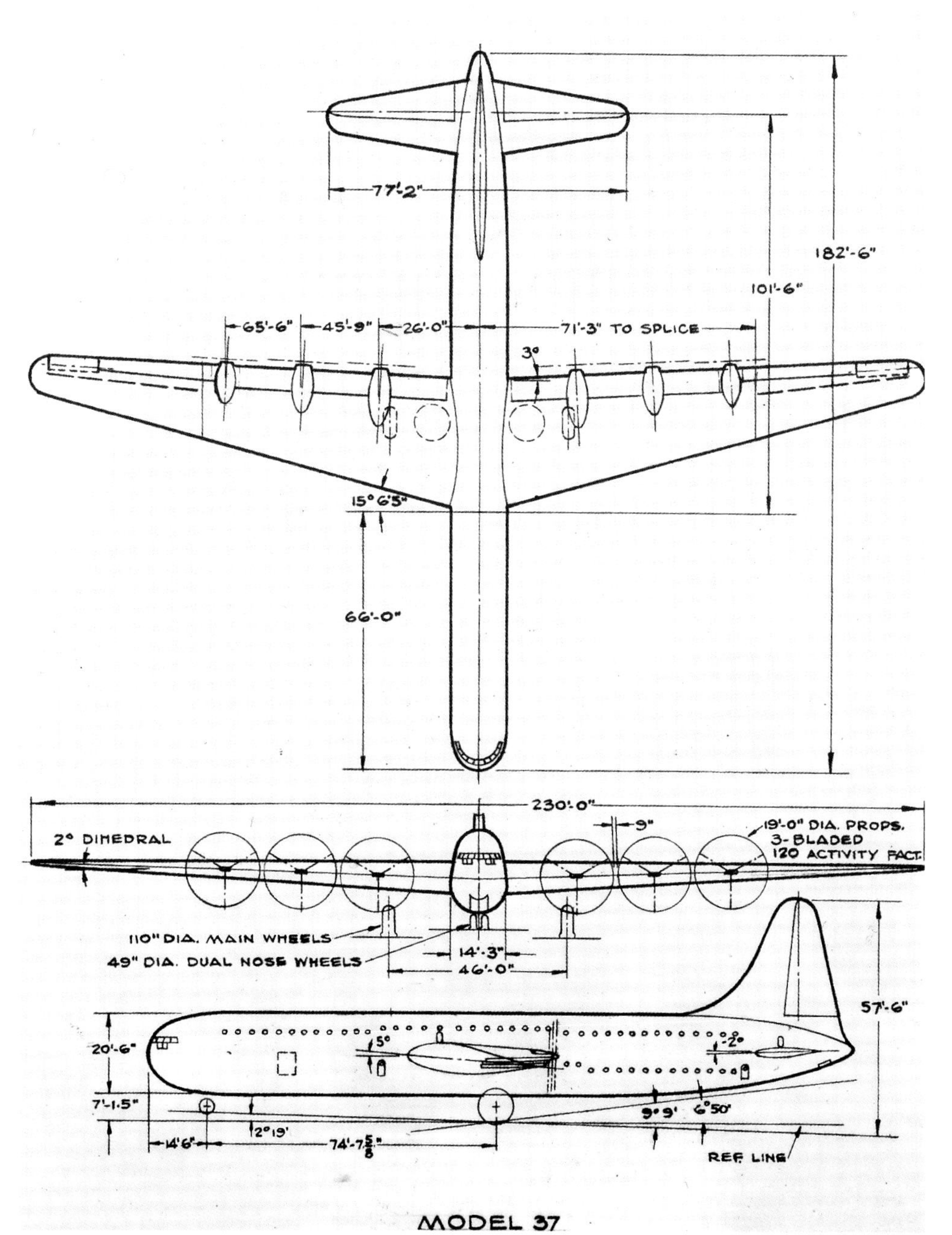

The XC-99, 43-52436, taking off from San Diego's Lindbergh Field in December 1947. It first flew, for nearly three hours, on November 24th—piloted by Russell R. Rogers and Beryl Erickson, the XB-36's original pilot. Notice the XC-99 used the same single wheel main landing gear as the XB and YB-36 prototypes. (Convair)

3-view drawing of Model 37/XC-99's configuration. Basic dimensions were: wingspan 230 ft. (same as the B-36), fuselage 182 ft. 6 in, fuselage height 20 ft. 6 in, height of tail 57 ft. 6 in. At the time it was the largest landplane in the world, a contemporary of the Hughes Flying Boat—the largest seaplane ever built. Not until the appearance of the C-5A and the 747 did the XC-99 shrink a little in size. (San Diego Aerospace Museum)

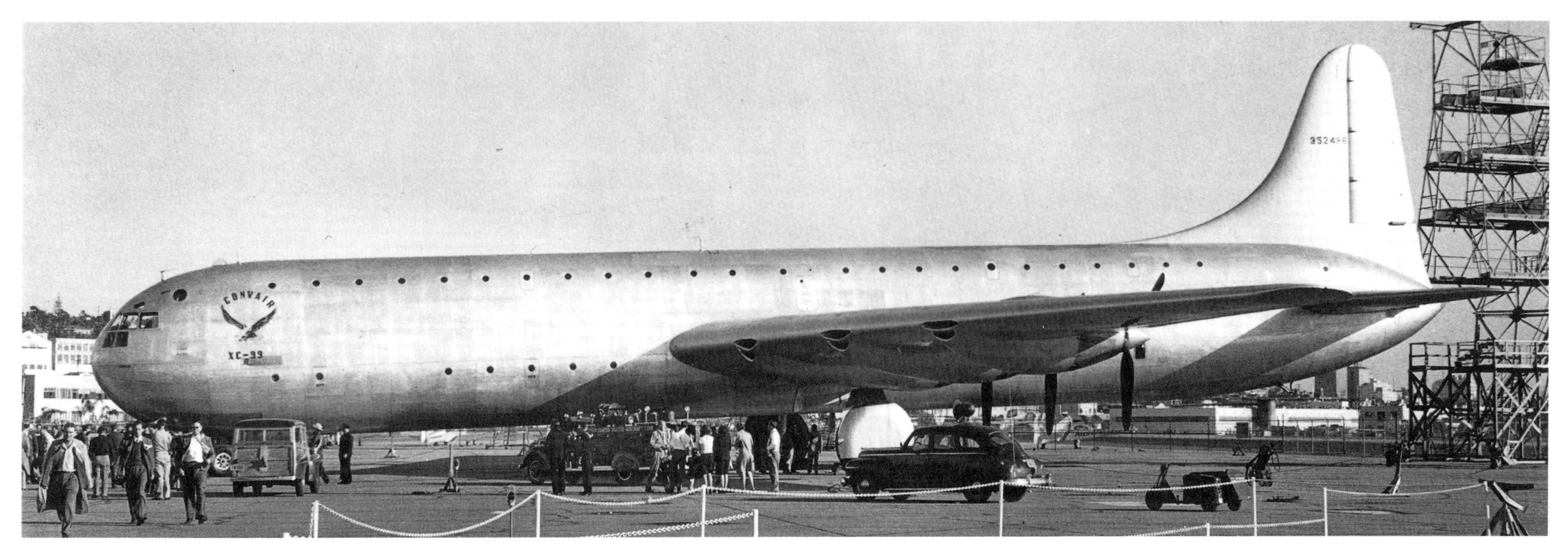

Finished XC-99 parked temporarily on an apron at Lindbergh Field near the Convair plant. Notice the Convair-owned fire truck kept close in case of an emergency. Late summer 1947. (Rohr Aircraft/San Diego Aerospace Museum)

Clean lines of the XC-99 are evident in this photograph taken over the sea off the southern California coast. (Convair)

In flight with its new four wheel main landing gear extended, the XC-99 prepares to land. Like the B-36, the multi wheel gear permitted operation from many more airfields. The plane first flew with this gear in January 1949. (San Diego Aerospace Museum)

Flight deck of the XC-99. An Air Force crew usually was eight men, including two rear scanners to monitor powerplants in flight. Notice position of the flight engineer and large center control console that could be reached by him. (Edwards AFB Office of History)

The XC-99 near the end of its glory days in 1955. A radome has been added to the nose and white anti-thermal paint to the top of the nose section to reduce heat. During the 7,400 hours that the XC-99 was flown, it broke 21 international records for cargo-carrying aircraft. Though the XC-99's record was impressive, it never achieved a production contract from the Air Force. After nearly eight years of service operating out of Kelly AFB, San Antonio, Texas, the XC-99 was retired in March 1957. Today, the XC-99 is stored at Kelly AFB, needing restoration work and a new museum home. (Author's collection)

The YB-60 was an all-jet version of the standard B-36 bomber. It was an attempt to compete with the Boeing B-52 Stratofortress which was planned to be the Air Force's follow-on bomber to the B-36. Two B-36Fs, 49-2676 and 49-2684, were pulled off the assembly line and redesignated as B-36Gs. Both these planes were finished with sweptback wings, a new taller sweptback tail and a more tapered nose. However, the YB-60, as it was again redesignated, did share 72% parts commonality with the production B-36. But it was powered by eight Pratt & Whitney J57 turbojets, each developing 8,700 lbs. of static thrust, same as those installed on the B-52. Seen here is YB-60, 49-2676, lifting off at Convair Fort Worth for its initial flight on April 18, 1952—Beryl Erickson, the test pilot. This photograph was censored during this early Cold War period, so the landing gears have been blacked out due to security requirements. In reality, it used the same B-36 gear. (San Diego Aerospace Museum)

The YB-60 prototype in flight. It was, at the time, the largest jet airplane in the world. Boeing's Dash 80 and the XB-52 prototype were both smaller in size. Wingspan was 206 ft, fuselage length 171 ft. and tail height 60 ft. 5 in. Gross weight was 410,000 lbs, the same as the B-36J. Only the forward crew compartment was retained, allowing a crew of five: pilot, co-pilot, navigator, bombardier/radar operator, and radio operator/tail gunner. Bomb load was about the same as a B-36. Convair crews flew the YB-60 some 66 hours accumulated in 20 flights. (Convair)

The first YB-60, 49-2676, was flown to Edwards AFB's Flight Test Center in January 1953 for Phase II flight tests. The Air Force flew the plane for only 15 hours on four flights. The flight test program was abruptly canceled on January 20th. The performance of the YB-60 was disappointing; top speed was 508 mph, some 100 mph slower than the B-52. The second YB-60 never received its engines and was scrapped without ever being flown. Both planes were cut up at Convair in July 1954. (Edwards AFB Office of History)

The NB-36H was a modified B-36H (51-5712) that became the first airplane to carry an operating nuclear reactor in flight. An early step in trying to develop a nuclear-powered bomber for the Air Force, the experimental plane was used to gather information on the most effective ways to shield crews from the effects of radiation, and to gather data about the effects of radiation on various aircraft components. Shown here in July 1955, the NB-36H had a special leadshielded crew compartment with a windshield up to eleven inches thick made of transparent shielding materials. Notice the open heavy-shielded escape hatch on top of the crew compartment. (Convair/National Atomic Museum)

Operating mainly over remote, unpopulated areas of New Mexico and Texas, the reactor never actually powered the NB-36H. A closed-circuit television system monitored the reactor and the plane's ten engines in flight. A total of 47 flights were made by the plane, nicknamed the "Convair Crusader," from September 17, 1955, to March 28, 1957. Notice the circular radiation symbol on the tail, and the air scoop for reactor cooling on each side of the rear fuselage. (General Dynamics Convair)

The NB-36H designation, with the letter "N" prefix, did not stand for "nuclear," but rather for an aircraft permanently modified for test purposes, and after the tests were completed, it would not be returned to its original configuration. Exact disposition of the NB-36H is not known, but it was scrapped in 1957. Rumors of its burial in a Nevada mineshaft have recently surfaced, and the radioactive site is planned to be inspected by the Department of Energy as part of its cleanup plan. It will not be studied until after the year 2001 due to its low priority. (Convair)

One of the weaknesses of early jet fighters was their limited range and endurance. The FICON (FIghter CONveyor) project was an early 1950s attempt to extend the range of fighter and reconnaissance jets by having them operate as parasites from B-36 bombers designated GRB-36s. Here GRB-36D 49-2696 loads a RF-84F, 51-1847 which was the reconnaissance version of the Republic Thunderstreak—renamed Thunderflash. Those fighters modified with a latchhook for FICON were redesignated RF-84Ks. Operationally, the fighter most often rendezvoused with the mothership after take off. (ACME)

Ten GRB-36Ds with cradle mechanisms to receive RF-84Ks in flight were teamed up with 25 RF-84Ks from the 91st Strategic Reconnaissance Squadron at Larson AFB, Washington. The GRB-36Ds were also based in Washington at nearby Fairchild AFB with the 99th Strategic Reconnaissance Wing. On a typical mission, the GRB-36 carried the fighter out to a 2,810 mile radius and launched the parasite at an altitude of 25,000 ft. The RF-84K then could make a high-speed photo run. The parasite could be picked up in mid-air, or while enroute to the target area. Night operations were possible, but more risky. (General Dynamic Convair)

Close-up of the retrieving cradle on the prototype GRB-36F, 49-2707. Notice prototype YRF-84F, 49-2430 modified with a latch-hook and downturned tail surfaces. Once attached and partially retracted into the modified bomb bay, the pilot of a RF-84K could climb out of his plane and enter the GRB-36D in flight. He could then change film, cameras, refuel, or even use the toilet facilities. However, this teaming arrangement of the 99th SRW and 91st SRS was discontinued after less than a year of operations in mid-1956. Withdrawal of the GRB-36D/RF-84K composite coincided with the introduction of the Lockheed U-2 spyplane into service. (General Dynamics Convair)

Project TOM-TOM was another approach to the parasite fighter program started after FICON had begun. F-84s were to be towed by means of wingtip hook-up attachments. Some of the advantages were: to provide improved penetration into a target area; being able to strike multiple targets; and being able to place more bombs on a single specified target. Shown here is a RF-84F making a successful wingtip docking with GRB-36F, 49-2707. However, the inherent danger involved in coupling the two aircraft eventually led to the cancellation of the program in 1953. (Convair/Dave Menard)

Probably the most distinctive feature of the B-36 Peacemaker was its massive size, dwarfing all other production aircraft of its era. Consequently, it was easy to see why the B-36 was often compared in publicity photos posed with a smaller plane to dramatically show off the Air Force's giant new bomber. Here in the north yard at Convair Fort Worth, the XB-36 displays its unique nose profile in comparison with a civilian lightplane. With wings removed, several such lightplanes could easily be accomodated in the B-36's bomb bays. (David Ånderton)

On a test flight in fall 1946, the XB-36 is accompanied by a company chase plane, a Convair C-87. This C-87 was a transport version of the famous B-24 Liberator heavy bomber. Wing span of the C-87 was 110 ft, about the size of just one of the B-36's huge wings. (USAF)

One giant shrinks to another! The XB-36 prototype is positioned next to the Air Force's biggest bomber of WWII. A B-29 had been taxied over from Carswell Air Force Base just for this publicity photograph. (Convair)

Another lightplane, looking like a toy, rests under a wing of B-36B, 44-92039, at a 1948 airshow. (David Menard)

Special group photo of Air Force bombers from the 1930s through the 1950s: a Douglas B-18 "Bolo," a Boeing B-17 "Flying Fortress," a Boeing B-29 "Superfortress," and the newest bomber, the B-36 "Peacemaker" dominating the group portrait with a 230 ft. wingspan. (USAF)

B-36 WINGS AND SQUADRONS

5th Strategic Reconnaissance Wing (later Bomb Wing), Travis AFB, California.
Squadrons were the 23rd, 31st and 72nd Bomb Squadron. January 9, 1951, to September 30, 1958.
TAIL CODE: CIRCLE X. Circle was 15th Air Force, X identified the 5th SRW/BW.

6th Bomb Wing, Walker AFB, New Mexico.
Squadrons were the 24th, 39th and 40th Bomb Squadron. August 28, 1952 to August 27, 1957.
TAIL CODE: TRIANGLE R (unverified). Triangle was 8th Air Force, R identified the 6th BW.

7th Bomb Wing, Carswell AFB, Texas.
Squadrons were the 9th, 436th and 492nd Bomb Squadron. June 26, 1948 to May 30, 1958.
TAIL CODE: TRIANGLE J. Triangle was 8th Air Force, J identified the 7th Bomb Wing.

11th Bomb Wing, Carswell AFB, Texas.
Squadrons were the 26th, 42nd and 98th Bomb Squadon December 1, 1948 to December 13, 1957.
TAIL CODE: TRIANGLE U. Triangle was 8th Air Force, U identified the 11th Bomb Wing.

28th Strategic Reconnaissance Wing (later Bomb Wing), Ellsworth AFB, South Dakota.
Squadrons were the 72nd, 717th and 718th Bomb Squadron. July 13, 1949 to May 29, 1957.
TAIL CODE: TRIANGLE S. Triangle was 8th Air Force, S identified the 28th SRW/BW.

42nd Bomb Wing, Loring AFB, Maine.
Squadrons were the 69th, 70th and 75th Bomb Squadron. April 1, 1953 to September 15, 1956.
TAIL CODE: No tail code assigned, as the code was phased out in 1953.

72nd Strategic Reconnaissance Wing (later Bomb Wing), Ramey AFB, Puerto Rico.
Squadrons were the 60th, 73rd and 301st Bomb Squadron. October 27, 1952 to January 1, 1959.
TAIL CODE: SQUARE F. Square was 2nd Air Force, F identified the 72nd SRW/BW.

92nd Bomb Wing, Fairchild AFB, Washington.
Squadrons were the 325th, 326th and 327th Bomb Squadron. July 29, 1951 to March 25, 1956.
TAIL CODE: CIRCLE W. Circle was 15th Air Force, W identified 92nd Bomb Wing.

95th Bomb Wing, Biggs AFB, Texas.
Squadrons were 334th, 335th and 336th Bomb Squadron. August 31, 1953 to February 12, 1959.
TAIL CODE: No tail code assigned as the code was phased out in 1953.

99th Strategic Reconnaissance Wing (later Bomb Wing), Fairchild AFB, Washington
Squadrons were the 346th, 347th and 348th Bomb Squadron. August 1, 1951 to September 4,
1956. TAIL CODE: CIRCLE I Circle was 15th Air Force, I identified the 99th SRW/BW.

B-36 SERIAL NUMBERS AND PRODUCTION QUANITIES

Serial numbers	Quantity	Model	Remarks
42-13570	1	XB-36	Prototype
42-13571	1	YB-36	Modified to RB-36E standard.
44-92004/92025	22	B-36A	All but 004 modified to RB-36E.
44-92026/92098	73	B-36B	
49-2647/2668	22	B-36D	
49-2669/2675	7	B-36F	
49-2676	1	YB-36G	Later redesignated YB-60.
49-2677/2683	7	B-36F	
49-2684	1	YB 36G	Later redesignated YB-60.
49-2685	1	B-36F	
49-2686/2702	17	RB-36D	
49-2703/2721	19	RB-36F	
50-1064/1082	19	B-36F	
50-1083/1097	15	B-36H	
50-1098/1102	5	RB-36F	
50-1103/1110	8	RB-36H	
51-5699/5742	44	B-36H	51-5712 redesignated NB-36H
51-5743/5756	14	RB-36H	
51-13717/13741	25	RB-36H	
52-1343/1366	24	B-36H	
52-1367/1392	26	RB-36H	
52-2210/2226	17	B-36J	
52-2812/2827	16	B-36J	
TOTAL	385 aircraft		

B-36 AIRCRAFT LOSSES DUE TO ACCIDENTS AND CRASHES

Like all military aircraft, the B-36 had its share of accidents and losses. Normal development problems and the demand of SAC operations took their toll. By the end of the ten year service life of the Peacemaker, 32 planes had been destroyed in various mishaps, 22 of which were flying accidents or crashes. Tragically, 176 officers and crewmen lost their lives.

B-36 accidents should not be singled out for special attention, since other bomber programs that were contemporary with the B-36 had their own tragic losses. Considering the large amount of flying time accumulated by the B-36 during training missions that could be 20 to 40 hours in length, the loss rate per hour flown was certainly better than average. Most B-36 pilots and crewmen considered the B-36 a very safe airplane to fly.

To a large extent, the number of accidents per B-36 unit reflected the length of time the unit was active and when it began operations. The 7th and 11th Bomb Wings at Carswell AFB had the B-36 for almost a decade and were the first units to get the plane, putting it through its earliest trials. They consequently had a greater share of mishaps. At the other end, the 72nd SRW/BW at Ramey AFB in Puerto Rico, active for less than 7 years, only had a single ground accident loss. Notably, the 99th SRW/BW at Fairchild AFB, active for less than 6 years, was the only wing to never have a major accident.